Pierre Bourdieu

Modern European Thinkers

Series Editor: Professor Keith Reader,
University of Newcastle upon Tyne

The Modern European Thinkers series offers low-priced
introductions for students and other readers to the ideas and
work of key cultural and political thinkers of the post-war era.

Jean Baudrillard
Mike Gane

Edgar Morin
Myron Kofman

Walter Benjamin
Esther Leslie

André Gorz
Conrad Lodziak and Jeremy Tatman

Gilles Deleuze
John Marks

Guy Hocquenghem
Bill Marshall

George Bataille
Benjamin Noys

Régis Debray
Keith Reader

Julia Kristeva
Anne-Marie Smith

Pierre Bourdieu

A Critical Introduction

Jeremy F. Lane

Pluto Press

LONDON • STERLING, VIRGINIA

First published 2000 by Pluto Press
345 Archway Road, London N6 5AA
and 22883 Quicksilver Drive,
Sterling, VA 20166–2012, USA

www.plutobooks.com

British Library Cataloguing in Publication Data
A catalogue record for this book is available from the British Library

ISBN 0 7453 1506 2 hbk; 0 7453 1501 1 pbk

Library of Congress Cataloging in Publication Data applied for

Designed and produced for Pluto Press by Chase Production Services
Typeset from disk by Stanford DTP Services, Northampton
Printed in the European Union by Antony Rowe Ltd, Chippenham

Contents

For Ben Taylor, who first encouraged me
to think about French culture.

Acknowledgements

The PhD thesis on which this study is based was born of a snap decision to abandon a lucrative job at a London radio station, move to Glasgow and become a student once again. Pierre Bourdieu notes somewhere that those who have a decent stock of inherited cultural and economic capital can easily afford to make the most apparently risky career moves and this was no doubt true in my case. Nonetheless, the social and psychological risks involved in such a move could have been real enough, had it not been for the constant support of friends, family and colleagues in London and the immense warmth of my reception in the marvellous city of Glasgow. The individuals concerned are too numerous to be mentioned here by name, but they know who they are and without their support this project would never have come to fruition.

First amongst those who made a more tangible contribution to the completion of this study is my thesis supervisor, Professor Sian Reynolds, without whose acute mind, encyclopedic knowledge of French history and society, and unfailing good humour my work on Bourdieu would not have progressed far. Of the many other individuals who contributed their time, advice and encouragement, I should also like to thank Jon Beasley-Murray, Pierre Bourdieu, Peter Collier, Martyn Cornick, Jill Forbes, Michael Kelly, Bill Kidd, Bill Marshall, Martin McQuillan, Robin Purves, Graeme MacDonald, Graham Roberts, Stephen Thomson and all my colleagues in the French Departments at the Universities of Stirling and Aberdeen. The Faculty Research Committee of the University of Aberdeen offered much appreciated financial help in covering copyright expenses. The editorial board of *Modern and Contemporary France* kindly agreed to my reproducing, in Chapter 5, a modified version of an article, 'Domestiquer l'exotique, exotiser le domestique: the symbiosis of ethnology and sociology in the work of Pierre Bourdieu', that first appeared in their journal in 1997, vol. 5, no. 4, pp. 445–56.

Introduction

In late August 1998, a photograph of the sociologist and anthropologist Pierre Bourdieu appeared on the front cover of the weekly French news magazine *L'Evénement du jeudi*. Below the photograph was a caption which read: 'Bourdieu, the most powerful intellectual in France'. Inside the magazine, a lengthy 'dossier' discussed the perceived strengths and weaknesses of Bourdieu's work.[1] That same month, similar 'dossiers' dedicated to Bourdieu appeared in the other principal French weeklies and dailies, *Le Nouvel observateur*, *Libération*, *L'Express* and *Le Monde*.[2] As the satirical French weekly *Le Canard enchaîné* put it, 'Bourdieumania' seemed to have broken out in the French media.[3]

No doubt Bourdieu himself was less than delighted with this sudden burst of media attention, sparked as it was by the publication of Janine Verdès-Leroux's highly critical study of his work, *Le Savant et la politique: essai sur le terrorisme sociologique de Pierre Bourdieu* (Politics and the Scholar: an essay on Pierre Bourdieu's sociological terrorism) (see Verdès-Leroux 1998). Nonetheless, the intense media interest in Bourdieu's work did genuinely reflect his current status as one of France's most high-profile intellectuals, a status gained both through his extensive theoretical output in the domains of sociology and anthropology and his increasingly frequent political interventions in defence of France's most impoverished and marginalised social groups, the homeless, the unemployed, illegal immigrants, striking workers. Bourdieu's name perhaps first began to be associated with these social and political issues in 1993, with the publication of the immense collaborative study of contemporary forms of social exclusion published under his editorship, *The Weight of the World* (*La Misère du monde* 1993). The book clearly resonated with a broader sense of political, cultural, and social malaise in France; it sold in large numbers and certain of its passages have been performed as short plays by French theatre groups. Bourdieu's public intervention in support of striking students and workers during the French public sector strikes of autumn 1995 raised his political profile still further. It was surely this experience of political engagement which encouraged him to found and edit the *Raisons d'agir* series, a set of pamphlets presenting 'the most advanced state of research on current political and social problems' written by a

1

range of authors 'all animated by the politically engaged desire to circulate the knowledge indispensable for political reflection and action in a democracy' (Halimi 1997, p. 4). Bourdieu has himself published two pamphlets in this series; an analysis of the threat to intellectual and artistic autonomy posed by the media in *On Television and the Media* (*Sur la télévision* 1996), and a series of directly political interventions, 'propositions for resisting the neo-liberal invasion', in *Acts of Resistance* (*Contre-feux* 1998). Both books have prompted further heated debate and sometimes quite hostile responses in the French media (see Schneidermann 1999).

Bourdieu's fame and influence are, however, by no means solely attributable to his recent, more directly political pronouncements in his native France. On the contrary, his work has for many years now exerted a significant influence over the fields of sociology and anthropology throughout the world. His theories of class, culture, and education, elaborated in texts such as *The Inheritors* (*Les Héritiers* 1964), *Reproduction* (*La Reproduction* 1970), and *Distinction* (*La Distinction* 1979), are obligatory points of reference for anyone writing in these areas. His three book-length studies of Kabylia in Algeria, *Esquisse d'une théorie de la pratique* (1972), its much revised and extended English translation, *Outline of a Theory of Practice* (1977), and *The Logic of Practice* (*Le Sens pratique* 1980), have acquired a similar status within the domain of anthropology. The publication in 1992 of Bourdieu's detailed study of the structure and genesis of the nineteenth-century French artistic and literary fields, *The Rules of Art* (*Les Règles de l'art* 1992), meanwhile, has extended his influence in the areas of literary and cultural studies. The year 1997, alone, saw the publication of two introductions to Bourdieu's work in French and two monographs in English, a special number of the academic journal *Modern Language Quarterly*, and two international conferences, at the universities of Southampton and Glasgow, dedicated to his work (Accardo 1997; Bonnewitz 1997; Fowler 1997; Swartz 1997; Moi, ed. 1997).

The immense public interest surrounding Bourdieu's political interventions, combined with the academic interest in his more specialised theoretical output, as manifest in the huge secondary literature generated by his work and the rapidity with which his new works are translated, suggest that his status as a significant international thinker is beyond question. However, if a broad consensus exists regarding the importance of Bourdieu's thought, the same cannot be said when it comes to situating that thought with regard to other intellectual traditions and movements. As the authors of the volume *An Introduction to the Work of Pierre Bourdieu* (1990) have pointed out, Bourdieu 'has been authoritatively placed in all major theoretical traditions', Marxist, Weberian, Durkheimian, even 'poststructuralist' or 'postmodernist' (Harker et al. 1990,

p. 213). Loïc Wacquant has similarly emphasised what he terms the 'blurred visions', 'conflicting reactions', and 'fragmented readings' which have hampered Bourdieu's reception in the English-speaking world (in Calhoun et al. 1993, pp. 235–62). As Wacquant argues, such confusions can largely be attributed to a failure by critics adequately to grasp the range of theoretical traditions with which Bourdieu engages, that 'nexus of antagonistic and competing positions within which and against which Bourdieu developed his own stance' (p. 245). Clearly then, there is a need for a study which positions Bourdieu's work much more clearly in terms of the contemporary French and, by extension, international intellectual field. However, if the ultimate aim of such a study were to be simply to append some definitive classification to Bourdieu's work, to decide once and for all whether he was 'really' a Durkheimian, Weberian, Marxist, or even 'postmodernist', then this would prove a peculiarly academic, not to say ultimately rather futile exercise. As Bourdieu himself has frequently emphasised, his concepts emerged as tools for thinking about specific sociological problems. They were never intended as a set of self-sufficient theoretical entities, pretexts for dry conceptual exegesis. As he put it in *Méditations pascaliennes* (1997):

> the most striking of misunderstandings [regarding my work] comes from the fact that the reading of the *lector* is an end in itself, that such a reading is interested in texts, and in the theories, methods, or concepts those texts mobilise, not in order to do something with them ... but in order to gloss them by relating them to other texts (under the cover sometimes of epistemology, sometimes of methodology). This kind of reading conceals what is essential, namely not only the problems that the concepts proposed intended to name and resolve – understanding a ritual, explaining variations as regards credit, savings, or fertility, accounting for differential rates of academic success or museum visiting – but also the space of theoretical and methodological possibles which have allowed such problems to be raised, at that moment in time, and in those terms ... and which it is essential to reconstruct by means of historical study. (1997, p. 77)

Here, Bourdieu alluded to two different, but potentially interrelated historical contexts, which he saw as essential to an understanding of his work. Firstly, he referred to his own concept of the 'intellectual field', that structured space of competing, often antagonistic positions, 'the space of theoretical and methodological possibles', within which all intellectuals necessarily take a position whenever they speak or write on a particular issue. According to Bourdieu, understanding his own or any thinker's work necessitates grasping the structure and historical genesis of the intellectual field into which that work intervenes, that 'nexus of antagonistic and competing positions within and against which Bourdieu developed his own stance', to quote Wacquant again. Grasping the coordinates of the French intellectual field out of which Bourdieu's work emerged is surely particularly important. For Bourdieu is an extremely combative thinker; his approach to any of the subjects he covers has typically emerged out of a polemical exchange with other commentators on the same subject. The extent to which his own 'theory of practice', elaborated primarily in his works of Kabyle anthropology, emerged out of a critical reaction to both Jean-Paul Sartre's existential phenomenology and Claude Lévi-Strauss's structural anthropology is made quite explicit by Bourdieu and has elicited much commentary. Frequently, however, Bourdieu does not identify by name those in opposition to whom he develops his own approach, preferring instead to use a series of euphemisms which, if easily decipherable by those working within the Parisian intellectual field, are much less so for Anglophone critics with no specific knowledge of that field.

The second, related context to which Bourdieu alluded in the quotation above was the rather broader historical one from which his work has emerged, namely the specific problems which his theoretical concepts have sought to name and analyse. Expressed somewhat differently, this is to highlight the importance of understanding the social, historical, and cultural developments in postwar French society which Bourdieu's work has sought to analyse.

Perhaps the most frequent criticism made of Bourdieu's work concerns its perceived determinism and consequent inability to account for significant historical change. Even a favourable critic such as Bridget Fowler concludes that 'apart from his studies in decolonisation', Bourdieu 'has never undertaken' any 'protracted discussion of *transformation*' in the social, cultural, or political spheres (1997, p. 5). If this were true, then Bourdieu would have to be considered a very poor sociologist indeed, for his detailed studies of education, class, and culture in postwar France have coincided with a series of dramatic and rapid changes in these domains. The period from 1958 to the present day, the period which spans Bourdieu's publishing career, has seen a series of changes which have transformed the nature of French culture and society, changes which include the violent and traumatic process of decolonisation; 'les trente glorieuses', that thirty-year span of rapid postwar economic reconstruction and growth; the massive and unprecedented expansion in the higher education sector from the late 1950s onwards, whose effects were to be most strikingly manifested in the disturbances of May 1968; the gradual waning of the Left Marxist or *marxisant* project, which had held such sway over the postwar intellectual and political fields, in the face of the emergence of a peculiarly French form of neo-liberal discourse mitigated by and mediated through a long tradition of centralised state intervention.

Such changes in French society can be read as symptomatic of what Ernest Mandel (1975) has described as the shift from an 'imperialist' stage to a 'late capitalist' stage in capitalist accumulation, the shift undergone by all the major Western economies in the postwar period. Mandel argues that late capitalism is characterised by a combination of factors; decolonisation and the consequent shift from colonial to neo-colonial relations between the developed and underdeveloped economies; the mass entry of women into the workforce; the emergence of the multinational corporation; the growth of the tertiary or services sector and the advent of a 'consumer society', as the developed economies seek surplus-profits through the sale of an ever-increasing range of finished consumer goods to Western consumers. He suggests that the search for new markets and the need for constant technological innovation to provide this ever-expanding range of new products led in turn to the emergence of new strata of middle-ranking executives in marketing, advertising, and research and development functions, executives educated in the rapidly expanding higher education sector. Faith in the ability of technological change and economic growth to ensure prosperity for all, meanwhile, encouraged the emergence of certain technocratic, managerialist, and ultimately neo-liberal discourses.

Although Bourdieu himself has never used the term, his work, far from being intrinsically resistant to social change, has frequently taken as its subject matter the very shifts which Mandel sees as symptomatic of the transition to late capitalism. Indeed, Bourdieu himself has often been a key protagonist in the intellectual and political debates such changes generated, analysing the traumas of French decolonisation in his early work on Algeria in the 1950s and 1960s, commenting on the postwar expansion in French universities in *The Inheritors* and *Reproduction*, tracing the fallout of the events of May 1968 in *Distinction*, sketching the emergence of a technocratic elite imbued with a peculiarly French form of neo-liberal ideology in *The State Nobility* (*La Noblesse d'état* 1989), to give but a few examples. Even the three works of Kabyle anthropology, *Esquisse ...*, *Outline ...* , and *The Logic of Practice*, which ostensibly appear least concerned with the dynamics of social change, can be understood in the context of late capitalism and of the effect that decolonisation has had in prompting a wholesale re-thinking of the forms and procedures of classical anthropology, a sustained reflection on the relationship between colonial power and the practice of anthropology. In all three studies, Bourdieu manifests an exemplary concern to think through the implications of this relationship between colonial power and anthropological knowledge and hence avoid the distortions implicit in the 'objectifying' gaze adopted by Western anthropologists on the distant, 'primitive' or 'exotic' societies they have typically studied.

It is important to stress that to argue that Bourdieu's work is best understood as an attempt to make sense of France's transition to late capitalism is not the same as arguing that Bourdieu is a 'postmodernist'. In a highly influential study, Frederic Jameson (1991) has argued that 'postmodernism' is 'the cultural logic of late capitalism'. Further, several critics have justified their claims that Bourdieu is himself a 'postmodernist' by pointing to his emphasis, in works such as *Distinction*, on the way that class identities in late capitalism are defined by lifestyle and patterns of *consumption*, rather than by position in relation to the means of *production*, as under a classically 'modernist' Marxist analysis (Lash 1990; Featherstone 1991). Bourdieu himself, however, has always reacted with extreme hostility to the suggestion of any affinity between himself and those French intellectuals generally considered 'postmodernist', such as Jacques Derrida, Jean Baudrillard, Jean-François Lyotard, Gilles Deleuze or Michel Foucault.

Any assessment of the validity of claims that Bourdieu is essentially a 'postmodernist' thinker is rendered problematic by the notorious imprecision of the term, as of the adjective 'poststructuralist' with which it is often conflated and which has also been applied to Bourdieu's work. Sadly, there appear to be almost as many conflicting definitions of the two terms as there are books dedicated to the subject. Thus, assessments of Bourdieu's relationship to 'post-structuralism' or 'postmodernism' will vary widely depending upon the particular definitions of these terms critics choose to employ, so much so that it is questionable whether the two terms have any real explanatory or analytical value. However, given that critics continue to use them to situate, criticise, or praise Bourdieu's work, this study will be obliged to retain them. For the purposes of this study, then, 'poststructuralism' will be used to refer to a broad and diverse group of French thinkers, including Derrida, Lyotard, Deleuze, Baudrillard, and the later Foucault, whose work emerged, at least to some extent, as a critical reaction against the perceived limitations of the structuralist paradigm which had dominated the French intellectual field from the late 1950s onwards. What these poststructuralists share, in spite of their many differences, and what has ensured their work is often also characterised as 'postmodernist', is a concern to challenge the classically modernist conceptions of subjectivity, Reason, Science, and History, to question the self-identity of the rational subject in the name of a structure of desire, difference, or alterity which cannot be dialectically 'sublated' (negated, preserved and transcended) into a modernist meta-narrative of History or Reason. Significantly, it is precisely the challenge to classical conceptions of Science, Reason, and History implicit in 'the postmodern turn' that Bourdieu has so trenchantly criticised, arguing it represents a 'thinly veiled nihilistic relativism ... that stands at the polar opposite to a truly reflexive social science' (1992, p. 52 [p. 72]). Indeed, given that Bourdieu himself has been at pains to emphasise his distance from what he has termed 'the nihilistic attack on science', typical of 'certain so-called "postmodern" analyses' (1984, p. 291 [p. xii]), there can be little justification for claiming him as a 'postmodernist' or 'post-structuralist'.

Paradoxically, Jameson's contention that 'postmodernism' is the 'cultural logic of late capitalism' rests on the very conception of History, the very historicist assumptions which those theorists most frequently dubbed 'postmodernist' have sought to question or deconstruct. To argue that a change in the nature of capitalist accumulation, such as the advent of late capitalism, finds its expression in a set of 'postmodern' cultural, social, and political forms is to rely on a model of 'expressive causality' which can have no place in a genuinely postmodern theory of culture, history and society, always supposing such a thing were possible.[4] This study will, therefore, distinguish between 'late capitalism', understood as an eminently *modernist* tool for theorising the development of postwar Western capitalism, and 'postmodernism', understood as a form of thought which would seek to challenge the assumptions implicit in the very notion of late capitalism.

In the chapters which follow, then, 'postmodernism' and 'post-structuralism' will be understood as referring to two of the competing positions structuring the contemporary French intellectual field, two of the various intellectual currents in opposition to which Bourdieu elaborated his own thought. 'Late capitalism', on the other hand, will be understood as referring to the broader set of social and cultural changes which have formed the subject matter of so much of Bourdieu's work. Situating Bourdieu's work in the two interrelated contexts of the French intellectual field out of which it emerged and the shift to late capitalism which it has analysed, this study will follow a broadly chronological and thematic approach, tracing the development of Bourdieu's thought from his earliest to his most recent works. Such an approach cannot claim to be exhaustive, to comment authoritatively on the entirety of Bourdieu's huge and varied output. Works of Bourdieu's which are less directly addressed to the issues contingent on the transition to late capitalism, such as his study *Language and Symbolic Power* (*Ce que parler veut dire* 1982) or his analysis of *The Political Ontology of Martin Heidegger* (*L'Ontologie politique de Martin Heidegger* 1988), will of necessity receive less detailed treatment. This study will nonetheless provide a detailed introduction to and analysis of all of Bourdieu's key concepts and theoretical approaches.

Finally, to insist on the importance of understanding the historical and intellectual context in which Bourdieu has elaborated his sociological theories is not to suggest that the significance of those theories is limited to that specific context. Bourdieu himself has frequently emphasised that his theories have pretensions to a general, 'transhistorical' validity. However, he has also emphasised that to arrive at this level of generality it is necessary 'to immerse oneself in the particularity of an empirical reality, historically situated and dated ... in order to construct it as a "particular case of the possible", as Gaston Bachelard put it, that is to say as an exemplary case in a finite universe of possible configurations' (1994a, p. 16 [p. 2]). Focusing on the particular empirical or historical realities which Bourdieu has sought to analyse might thus be seen as a necessary precursor to any objective assessment of the general validity of his ideas or concepts. In other words, it is only by understanding the specificity of the field of French higher education in the 1960s, for example, that we can decide on what conditions, at the price of what modifications, Bourdieu's conclusions regarding education, class and culture might be relevant to our own national experience and historical context.

Notes to the Introduction

1. See *L'Evénement du jeudi*, no. 721 (27 août – 2 septembre 1998).
2. See *L'Express* (20 août 1998), *Libération* (28 août 1998), *Le Monde des livres* (28 août 1998), *Le Nouvel observateur*, no.1765 (3–9 septembre 1998).
3. See 'Bon Dieu de Bourdieu', *Le Canard enchaîné* (2 septembre 1998).
4. For a more detailed analysis of the inconsistencies inherent in locating 'postmodernism' within a fundamentally modernist historical meta-narrative, see Bennington (1993, pp. 172–95), and Lyotard (1988, pp. 105–13).

CHAPTER 1

Peasants into Revolutionaries?

The broad details of Pierre Bourdieu's background and early intellectual career have been well documented, both in existing critical studies and in interviews. He was born in 1930, the son of a postman in a peasant community in the Béarn in the French Pyrenees. Having passed through the *classes préparatoires* at the renowned Lycée Louis-le-Grand in Paris, he entered the elite École normale supérieure to study for an *agrégation* in philosophy, perhaps the most prestigious academic qualification in France at that time. Bourdieu obtained his *agrégation* in 1954 but, frustrated by the abstract tenor of academic philosophy, he abandoned his plans to prepare a doctoral thesis under the supervision of Georges Canguilhem on the work of Maurice Merleau-Ponty and took up a post as a philosophy teacher in a provincial *lycée* instead. His career as a philosophy teacher was short-lived and in 1955 he was drafted into the French Army and sent to Algeria where he served as a conscript during the Algerian War of Independence. His military service completed, Bourdieu remained in Algeria, working as an *assistant* in the Faculté des lettres of Algiers University, whilst carrying out fieldwork in the country's cities, villages, and 'resettlement centres'. In 1960, he returned to France to become Raymond Aron's assistant at the Sorbonne.[1] On the basis of his fieldwork in Algeria, Bourdieu published a series of books and articles between 1958 and 1964 which examined the country's social, economic, and political development. As Bourdieu has put it, if he arrived in Algeria in 1955 a philosopher, by the time he left he had become a sociologist (Honneth et al. 1986).

Although Bourdieu's experiences in Algeria clearly represent a founding moment in his intellectual career, his early Algerian work has received relatively little critical attention to date. This is presumably because the work largely pre-dates the elaboration of such key Bourdieusian concepts as 'habitus', 'field', 'cultural capital' or 'symbolic violence', whilst its subject matter appears tangential to the analyses of class, culture and social reproduction upon which Bourdieu's mature reputation rests. Nonetheless, as this chapter will attempt to demonstrate, these early studies of Algeria are important for the insights they offer into the way Bourdieu theorises social and cultural change, insights which will prove vital to an understanding of his later work on socio-cultural change in French society. It was in the course of his work on the Algerian peasantry and sub-proletariat that Bourdieu was to lay the theoretical foundations of his later studies, anticipating in important respects concepts such as 'habitus', 'practice' and 'field'. On a more general level, Bourdieu's decision to abandon his philosophical studies in favour of a 'scientific' practice of sociology can tell us much about how he understands his own position within the French intellectual field.

Sociology over Philosophy

In subsequent accounts of his very early career, Bourdieu has suggested that his modest social background, his consequent sense of being an outsider in the Parisian intellectual field, and his frustration at the limitations of academic philosophy were all linked to his decision to undertake empirical sociological studies of Algeria and hence to his 'conversion' from philosopher into sociologist. To practise 'scientific' sociology, in Bourdieu's terms, demands achieving an objective distance on the hidden workings of the social world. To be an outsider is already to be endowed with such a distance, whilst to be an outsider who has nonetheless achieved academic success is to supplement that instinctive feeling of distance with a 'mastery of scientific culture'. Thus, he argues that 'to combine an advanced mastery of scientific culture' with 'a certain revolt against or distance from that culture (most often rooted in an estranged experience of the academic universe)' is to possess a socially determined disposition towards the practice of scientific sociology (Bourdieu 1992, p. 218 [p. 249]). As Bourdieu was to put it in an interview many years after the Algerian War:

> This more or less unhappy integration into the intellectual field may well have been the reason for my activity in Algeria. I could not be content with reading left-wing newspapers or signing petitions; I had to do something concrete, as a scientist ... . That's where my 'scientific bias' stems from. (in Honneth et al. 1986, p. 39)

Conducting empirical sociological research not only seemed to represent a more 'concrete' form of activity than pursuing the abstractions of academic philosophy, it also, Bourdieu has argued, offered the possibility of an 'objective', 'scientific' assessment of the situation in Algeria not available to those who commented from the sidelines. Whilst acknowledging the importance of the stand taken by intellectuals such as Jean-Paul Sartre, Francis Jeanson, and Pierre Vidal-Naquet 'against torture and for peace', he has explained that he was 'concerned about the associated utopianism' since 'it was not at all helpful, even for an independent Algeria, to feed a mythical conception of Algerian society'. Bourdieu has singled out Frantz Fanon's *The Wretched of the Earth* (*Les Damnés de la terre* 1961) for particular criticism, stating the book contained 'dangerous' analyses engendering 'a pernicious utopianism' amongst Algerian intellectuals. His own work of the period sought to challenge Fanon's assertion that the Algerian peasantry and sub-proletariat constituted a revolutionary force, a challenge supported by detailed empirical research, by 'observation and measurement, not through reflecting upon second-hand material'. Thus, the Algerian War was to find Bourdieu 'between camps' in terms of the French intellectual field, opposed to the continued French colonial presence in Algeria but critical of the analyses offered by some of those French intellectuals who supported the struggle for Algerian independence (in Honneth et al. 1986, pp. 38–40).

The account Bourdieu offers here of his conversion to sociology highlights his firm belief in the pre-eminence of an empirically-informed practice of 'scientific' sociology over what he sees as the abstract and ahistorical theorising of a particular form of French academic philosophy; 'the philosophical babble found in academic institutions' (1987, p. 13 [p. 5]). To recognise the importance of Bourdieu's account both of his own position and of the position of sociology in the intellectual field is not necessarily to accept that account unreservedly. Bourdieu clearly sees this account as a scientific objectification of his position in the intellectual field. However, it derives much of its force from an essentially *rhetorical* claim to greater personal authenticity and good faith, combined with an invocation of profoundly *subjective* feelings of alienation, which, however genuinely felt, surely form a questionable basis for a sociological practice which defines itself in terms of its *objectivity* and scientificity.

Bourdieu himself has warned of the dangers of conflating 'questions of science' with 'concerns of conscience' (1963a, p. 259). As he has pointed out, 'good intentions ... often make bad sociology' and assessments of any thinker's work, particularly in the politically charged context of colonialism, should focus not on questions of authenticity or sincerity, but rather on the adequacy of that thinker's 'problematic', on the 'concepts, methods and techniques' they employ (Bourdieu 1980, p. 14 [p. 5]). Whatever Bourdieu's motivations for becoming a sociologist, these cannot therefore be invoked in support of a claim to the a priori superiority of his analyses. Any assessment of the importance of his early work on Algeria must confine itself to an analysis of that work's inherent strengths and weaknesses.

Sociologie de l'Algérie

Bourdieu's first published book and his first detailed analysis of events in Algeria appeared in 1958 entitled *Sociologie de l'Algérie*. Echoing Max Weber's *The Methodology of the Social Sciences* (1949), Bourdieu emphasised in his Introduction that his approach was based on 'sober and objective observation', on 'disinterest and impartiality' (1958, p. 5). The first six chapters of the book were dedicated to a 'recon-struction' of 'the *original* social and economic structures' of Algeria's different indigenous ethnic groups, Kabyle, Shawia, Mozabite, and Arab. His analyses of these different societies were to be seen as '*ideal-types* in Max Weber's sense, the product solely of historical reconstruction – with all the uncertainties that this implies' (p. 90). This 'reconstruction' was seen as 'indispensable for understanding the phenomena of acculturation and deculturation determined by the colonial situation and the irruption of European civilisation', which he examined in the final chapter (p. 5). These phenomena of 'acculturation' and 'deculturation' were understood in terms of the incomplete imposition of 'modern' forms of economy and society, based on 'rational calculation' and 'capitalist accumulation', on to a fundamentally 'traditional' society. The problem, according to Bourdieu, was how best to ensure the transition between 'traditional' Algerian ways of life, now fatally undermined, and the 'modern' socio-economic forms introduced by the Europeans, without aggravating the phenomena of 'disintegration' already affecting Algeria's indigenous populations. Again in classically Weberian style, Bourdieu concluded that the objectivity of 'a properly sociological analysis' did not allow him to choose between the various solutions on offer, merely to exclude those that were objectively impractical (p. 126).[2]

Nonetheless, this 'properly sociological analysis' did allow Bourdieu to exclude 'certain options, such as "assimilation" and the maintenance of the colonial situation' (p. 126). As Bourdieu pointed out, French colonialism was an inherently contradictory enterprise; although it proclaimed its intention to 'civilise' and 'assimilate' its Algerian 'subjects' by ensuring their social and economic development, to succeed in such a project would undermine the hegemony of Algeria's European population and hence necessarily spell the end of the colonial regime. Thus, if colonialism had destroyed the traditional structures of Algerian tribal organisation, it could never construct a modern society in its place, he maintained, since this ran counter to the objective interests of the colonial regime.

Bourdieu's emphasis on the contradiction at the heart of the French colonial enterprise was an attempt to refute Germaine Tillion's analyses in her *L'Algérie en 1957* (1957). Tillion, an ethnographer and adviser to Jacques Soustelle, Governor-General of Algeria from 1955–56, had argued, on the basis of her research in the Aurès mountains, that the introduction of European medicines, famine relief, and the incursions of a money economy had all undermined the stability of traditional tribal society without being able to secure the tribespeople's integration into a 'modern', 'rational' economy. The Algerians of the Aurès had been abandoned 'in the middle of the ford', as she put it. If only the colonial authorities were to pursue an enlightened policy of aid and education, facilitating 'a true social mutation' amongst Algeria's indigenous populations, then the transition between 'tradition' and 'modernity' could be secured peacefully and the Algerian nationalists' grievances would disappear (Tillion 1957).

As Paul Clay Sorum (1977, pp. 84–8) has shown, Tillion's work was immensely influential amongst those he terms France's 'humanist', 'anti-colonialist' intellectuals, who were concerned with reforming the colonial system for the benefit of the colonised peoples rather than with ending the system itself. *Sociologie de l'Algérie* was an important reminder of the idealism and inconsistencies implicit in this position. As Bourdieu showed, if after more than a hundred years in Algeria the French had been unable to 'modernise' the structures of Algerian society, this was by no means accidental. Rather it reflected the objective logic of the colonial system, which demanded that the Algerians be kept in a permanent state of under-development. Bourdieu's refutation of Tillion's analysis was thus both an important contribution to debates in France about the Algerian situation and a significant gesture of solidarity with the Algerian independence movement. His attention, in the book's earlier chapters, to the richness and diversity of Algeria's indigenous cultures also represented an important riposte to the disparaging depiction of those cultures current amongst the more determined defenders of the colonial regime.

Bourdieu's analyses of 'traditional' Kabyle, Shawia, Mozabite, and Arab social structures owed much to the Durkheimian tradition. He suggested that there was an organic relationship between each group's religious, cultural, and economic practices and the characteristics of their natural environment. Social practices and natural environment formed the mutually determined and determining elements of an inherently stable entity, itself determined by the classically Durkheimian imperative of maintaining each group's 'solidarity', 'cohesion' and 'equilibrium'. Although noting numerous differences between these groups, Bourdieu argued that all were driven by the same inherent tendency to preserve their structures unchanged; they were characterised by 'social structures whose cohesion guarantees, as far as is possible, the equilibrium between man and the natural environment' (1958, p. 101).

In his *The Division of Labour in Society* (*De la division du travail social* 1893), Durkheim had identified Kabylia as a classical example of 'mechanical solidarity', of a society organised along 'segmentary principles', consisting of a structure of homologous, concentric, interlocking segments of family, clan and tribe, 'a series of successive interlockings [*une série d'emboîtements successifs*]' which ensured 'the unity of the total society' (1893, p. 153 [p. 178]). Bourdieu echoed this analysis, stating that Kabyle society, 'composed of successive interlocking collectivities [*comme par emboîtements successifs de collectivités*], presents a set of concentric allegiances' (1958, p. 13). Indeed, he argued that each of Algeria's 'traditional' societies was segmentary in the Durkheimian sense, marked by: 'A proliferation of undifferentiated cells, a simple juxtaposition of families, this strictly conservative society repeats and imitates the past rather than taking it up in order to move beyond it in a continuous progress, innovation being taken for an impure and impious magic which can only bring bad luck' (p. 101).

Notwithstanding his evident debt to Weber, Bourdieu's analysis in *Sociologie de l'Algérie* might be read as an exercise in what Durkheim called 'social morphology'. According to Durkheim's definition, social morphology involves the attempt to describe a given social or ethnic group's 'social substrate', that is to say, 'the mass of individuals, the manner in which they are arranged on their territory, the nature and configuration of the things of all sort which affect their collective relations'. Identifying a group's 'social substrate' was, for Durkheim, the necessary precursor to an analysis of that society's evolution, of the effects of gradual urbanisation, for example, or of changing population densities and distributions (Durkheim 1969, pp. 181–2). This, then, was Bourdieu's method in *Sociologie de l'Algérie*; identifying the original 'social substrate' of each of Algeria's ethnic groups before analysing the effects of colonialism and war thereon.

In accordance with this classically Durkheimian, 'social functionalist' problematic, Bourdieu understood any change to the static structures of 'traditional' Algerian society as being provoked by 'culture contact', the arrival, with colonisation, of a dynamic European civilisation which eroded the 'traditional' forms of Algerian culture and imposed 'modern' forms in their place in a process of 'deculturation' and 'acculturation'.[3] This was the process Bourdieu described in two consecutive chapters he contributed to the 1959 volume *Le Sous-développement en Algérie*, 'La Logique interne de la civilisation algérienne traditionnelle' (The Internal Logic of Traditional Algerian Society) and 'Le Choc des civilisations' (The Clash of Civilisations). He argued that a pre-disposition towards equilibrium and stasis formed 'the internal logic' of all of Algeria's indigenous tribal groups; in 'a traditionalist and customary social order, resistant to progress, turned towards the past', all effort was aimed at maintaining the existing equilibrium, so that 'the possibility of positing a better social order in reference to which the established order would be grasped as imperfect, is radically excluded' (1959a, pp. 50–1). Colonisation, 'the irruption of European civilisation', had shattered forever this state of 'traditional' equilibrium. The arrival of Europeans had alerted the Algerians to the existence of alternative forms of social, economic and cultural organisation; 'The European ... thus makes what seemed necessary appear as contingent, what appeared "natural" an object of choice' (1959b, p. 57).

Although refined and modified in subsequent years, this account of the changes affecting Algerian society might be seen as something of a model for Bourdieu's later theorisations of socio-cultural change in France. Thus, for example, in his analysis of the crisis in French universities in the 1960s in *Reproduction* and *Homo academicus*, Bourdieu would emphasise the importance of *morphological* change, of changes in the number, density, age, social background, and hence expectations of students and lecturers. It was, he would argue, the sudden arrival in the universities of these new kinds of student and lecturer, with their different values and expectations, which shattered the previously 'organic' stage in French higher education, in which the system's objective logic remained unquestioned because apparently 'natural', and replaced it with a 'critical' stage, in which everything that had previously been unquestioned became the subject of the sort of intense debate and argument that characterised the years around May 1968. This account of social change as constituting a temporary breakdown in an otherwise smoothly functioning system clearly owes more to Durkheim's notions of social 'cohesion' and 'equilibrium', to his distinction between 'normal' and 'pathological' social states, than it does to Marx's vision of societies as being riven by fundamental social antagonisms, 'between freeman and slave, patrician and plebeian, lord and serf, guild-master and journeyman, in a word, oppressor and oppressed' (Marx 1872, p. 79).

Where an orthodox Marxist might object in general terms to Bourdieu's reliance on Durkheimian notions of 'cohesion' and 'equilibrium', other objections might be raised regarding their specific application to Algeria's tribal societies. For Bourdieu's analyses seemed to rely upon a series of rather stark dichotomies between 'tradition' and 'modernity', the 'static' 'internal logic' of Algeria's social structures, inherently 'resistant to progress', and the 'irruption' of a destructive, but nonetheless dynamic 'European civilisation'. In the years following the Algerian War, numerous anthropologists and historians of North Africa have challenged such dichotomies, highlighting the dangers of exaggerating the apparent 'traditionalism' of tribal societies, whilst insisting on the need to understand such 'traditionalism' not, as the historian Abdullah Laroui puts it, as reflecting some fundamental disposition to stasis, but rather as 'the dialectical response to a historical blockage' caused by first Turkish then French colonial domination (1970, p. 64).[4]

Both Talal Asad (1973, pp. 103–18) and Abdelkebir Khatibi (1983) have questioned the validity of Durkheim's concept of segmentary societies and the whole social functionalist tradition in anthropology that flows from it. Both suggest that the social functionalist emphasis on equilibrium and cohesion reflects the vested interest that anthropologists from the Western colonial powers had in exaggerating the static, unchanging nature of 'traditional' societies in the colonies. Khatibi argues that the problem with Durkheimian accounts of 'segmentary' social structures and the equilibrium they are held to ensure is that they 'liquidate history through ignorance (?) of the evolution of the Maghreb, where centralised States have existed and intervened directly in tribal organisation. In this sense, the segmentary system is an epistemological illusion ... . It is difficult to see how this equilibrium maintains itself or escapes from the history of the society as a whole' (1983, pp. 100–1). This objection could equally be addressed to Bourdieu's analyses in *Sociologie de l'Algérie*. For, if 'traditional' Algerian society were indeed as static as Bourdieu suggested, as inherently resistant to initiative or innovation, it was difficult to see how he could account for the emergence of an indigenous liberation movement which was to prove powerful enough to provoke the downfall of a French Republic and the end of the French colonial presence in Algeria.

Indeed, it was striking that in this first edition of *Sociologie de l'Algérie* Bourdieu made no reference to the existence of an Algerian independence movement at all. Certainly, Bourdieu manifested a clear admiration for the initiative shown by indigenous Algerians in creating social structures perfectly adapted to their geographical surroundings. Further, he insisted that the unchanging nature of Kabyle society, for example, should be attributed to their *conscious* rejection of successive invasions, Roman, Turkish, French, and not to some inherent 'primitivism' (1958, pp. 16–17). Nonetheless, this seemed to attribute any conscious agency on the part of the Algerians to the imperative of continued social equilibrium; all genuine innovation, progress or change in Algeria, on the other hand, was attributed to colonisation, to the arrival of the Europeans.

But the emergence of Algerian nationalism could not be attributed simply to 'the irruption of European civilisation', to 'acculturation' and 'deculturation', if those terms are understood as referring to the unidirectional acquisition of 'modern' Western culture and the consequent shedding of indigenous 'tradition'. Algerian nationalism drew both on Western Marxism, which migrant Algerian labourers had first encountered in French factories between the wars, and indigenous traditions, notably Islam. To quote Benjamin Stora, 'situated at the intersection of two great projects, that of the socialist movement and that of the Islamic tradition', Algerian nationalism 'constituted a challenge to received ideas and categories and ideological a priori' (1992, p. 129). The emergence of Algerian nationalism might be better understood as a case of 'transculturation', a term coined by the Cuban anthropologist Fernando Ortiz (1940) to describe phenomena of cultural interchange, which are not reducible to the unidirectional process implicit in the notion of 'acculturation'.

Many of the objections raised here were to be answered in the more detailed and nuanced analyses of Algerian social, political, and economic development that Bourdieu would offer in his later studies *Travail et travailleurs en Algérie* (Work and Workers in Algeria) (1963) and *Le Déracinement* (Deracination) (1964). Moreover, in his subsequent articles, as in the various revised editions of *Sociologie de l'Algérie*, Bourdieu engaged more directly with the political aspects of the Algerian War, arguing first that the war had created the conditions for a wholesale socialist revolution in Algeria, before retreating from that position to express reservations about the prospects for a successful revolution. Before turning to a discussion of *Travail et travailleurs ...* and *Le Déracinement*, it will be useful briefly to trace the shifts in Bourdieu's political position.

From a Conflict to a War to a Revolution

A useful index of Bourdieu's increasing politicisation and radicalisation in the years which followed the publication of the first edition of *Sociologie de l'Algérie* can be found in the changing terminology he used to describe events in Algeria. Where previously he had referred to the events merely as 'a conflict' (1958, p. 125), in the 1960 article, 'Guerre et mutation sociale en Algérie' (War and Social Change in Algeria), he referred to them as a 'war' and identified the military wing of the *Front de libération nationale* (FLN), the *Armée de libération nationale* (ALN), as a significant force within Algeria. In the subsequent article, 'Révolution dans la révolution' (Revolution in the Revolution) (1961), he began to discuss the possibilities of the war of liberation transforming itself into a socialist 'revolution' in an independent Algeria. Such small changes of vocabulary were highly significant given the French authorities' persistent denials that the events in Algeria constituted a 'war', preferring instead 'a series of euphemisms such as "operations to maintain order" or "the events"' (Dine 1994, p. 7).

At the core of Bourdieu's argument in 'Guerre et mutation sociale en Algérie' and 'Révolution dans la révolution' was his contention that Algerian resistance to colonialism was entering a qualitatively different stage. Prior to the war, he argued, resistance to the French had largely taken the form of retrenchment, of a conscious adherence to tradition and a refusal of Western styles of dress, of Western medicines, and of cultural and social forms perceived to be the tools of the coloniser. However, as the war progressed, the Algerians were appropriating these tools of colonial domination to turn them against the French. Moreover, the war had overturned 'traditional' rigidly patriarchal familial structures to allow both women and young men to play a key role in the liberation struggle. The war was thus creating the objective conditions for a socialist revolution, the 'revolution in the revolution'. Bourdieu's analysis owed more than a little to Fanon's study *A Dying Colonialism* (*L'An V de la révolution algérienne* 1959), which had dealt with each of these points in turn a year previously.

Bourdieu's gradual radicalisation was reflected in two revised editions of *Sociologie de l'Algérie* published over this period. In 1961, a second edition of *Sociologie de l'Algérie* had appeared which included extended versions of the analyses Bourdieu had offered in the first edition and ended with a more determined call for 'the colonial system' to be 'radically destroyed, that is to say from top to bottom' in order that the Algerian masses should be able to 'assume their own destiny in complete freedom and full responsibility' and build a new society based 'on active, creative, and resolute participation in a common task' (1961, pp. 125–6). In 1963, a third edition was published, identical to the second except for its final sections in which Bourdieu argued that Algerian society, 'profoundly revolutionised by colonisation and war', demanded 'objectively, revolutionary solutions'. The mass exodus of the European population following independence in 1962 had made 'the accession to independence the chance for an economic and social revolution', demanding a choice 'between chaos and a form of socialism closely corresponding to the [Algerian] reality' (1963, p. 126). In the interim between the publication of the second and third editions of *Sociologie de l'Algérie*, its English translation, *The Algerians* (1962), had appeared, which concluded with an extended version of the article 'Révolution dans la révolution' which contained Bourdieu's most detailed analysis of the activities of the FLN, the extent of its support amongst the Algerian masses and its role in their politicisation.

If the articles 'Guerre et mutation sociale ... ' and 'Révolution dans la révolution' suggested a certain affinity between Bourdieu's analyses and those offered by Fanon, that affinity was not to last long. Moreover, the confidence expressed by Bourdieu in 'Révolution dans la révolution' that the war had so transformed 'traditional' Algerian society as to have provoked an inherently revolutionary situation was to be similarly short-lived. In the 1962 article, 'De la guerre révolutionnaire à la révolution' (From Revolutionary War to the Revolution), Bourdieu had already expressed uncertainties about the possibility of a successful transition to socialist revolution, whilst questioning Fanon's analysis in his 1961 *The Wretched of the Earth*. Fanon had argued that in the colonial context it was the peasantry and the sub-proletariat which constituted the primary revolutionary classes since the proletariat had a relatively privileged position within colonial society and thus a vested interest in its maintenance. Bourdieu replied that it was 'only at the price of an alteration of reality inspired by the desire to apply classical schemes of explanation that one can see the peasantry as the only revolutionary class' (1962a, p. 8). Whilst acknowledging the effect of the war in politicising the Algerian masses, he argued that their politicisation remained incomplete, leaving the Algerians 'floating' between two incompatible traditions, the indigenous and the Western. Such a situation called for a continuing programme of education to foster a coherent new political culture.

Bourdieu's increasing scepticism regarding the possibility of a successful socialist revolution in post-independence Algeria was to be reflected in one further rewrite of the final sections of *Sociologie de l'Algérie*. Thus, the edition currently in print concludes with a warning that to claim that the peasantry and sub-proletariat constitute a revolutionary force is to indulge in 'national populism', an ideology propagated by 'the new bourgeoisie of the great State bureaucracies', in an effort to secure the support of the masses for their political and economic hegemony (Bourdieu 1985, p. 125). Furthermore, Bourdieu has subsequently sought to distance himself from the 'naiveties' of his stance in articles such as 'Révolution dans la révolution' (1980, p. 8 [p. 2]). Bourdieu's final position on Algeria can be clarified by examining the series of articles he published between 1962 and 1963, which fed into the two book-length studies, *Travail et travailleurs en Algérie* (1963) and *Le Déracinement* (1964).

Temporal and Political Consciousness

In the series of essays and books which he published from 1962 onwards, Bourdieu was to take issue not only with the analyses contained in Fanon's *The Wretched of the Earth*, but also with the theory of revolutionary action elaborated by Sartre, Fanon's great ally, in his *The Problem of Method* (*Questions de méthode* 1960). The empirical basis for Bourdieu's critique was the fieldwork he had conducted between 1958 and 1961 amongst some three million

Algerian peasants who had either been forcibly re-located by the French authorities into ' resettlement centres' or who had fled the countryside in search of work and shelter in one of the shantytowns springing up on the outskirts of Algeria's cities. The importance of this work to Bourdieu's intellectual development was emphasised by his decision to republish, in slightly modified form, his contribution to the collaborative study *Travail et travailleurs en Algérie* in 1977 under the title *Algeria 1960 (Algérie 60* 1977).

Central to Bourdieu's dispute with Fanon and Sartre was the question of the distinction between Algeria's displaced peasants and sub-proletarians and a genuinely revolutionary proletariat. As Bourdieu was to put it later: 'I tried to show that the principle behind this difference is situated at the level of the economic conditions of possibility which allow for behaviour based on rational forecasting, of which revolutionary aspirations are one dimension.' He employed the ability to make rational calculations of future economic profit as a test of the revolutionary potential of these two social groups: 'I wanted also to understand, through my analyses of temporal consciousness, the conditions of the acquisition of the "capitalist" economic habitus among people brought up in a pre-capitalist universe' (1987, pp. 17–18 [pp. 7–8]).

It was in the three articles, 'La Hantise du chômage chez l'ouvrier algérien' (Fear of Unemployment amongst Algerian Workers) (1962), 'Les Sous-prolétaires algériens' (The Algerian Sub-Proletariat) (1962) and 'La Société traditionnelle: attitude à l'égard du temps et conduite économique' (Traditional Society: attitude to time and economic conduct) (1963), that Bourdieu fleshed out the critique of Fanon he had already sketched in 'De la guerre révolutionnaire ... '. He argued that whilst the war and the mass migrations from the countryside had eroded traditional values and social structures, the economic and cultural insecurity of the displaced peasants and sub-proletariat rendered them incapable of forming rational projects for the future. Forced to live from hand to mouth, the sub-proletariat had no means of making plans, of organising their present in terms of an abstract, still-to-be-realised future. Everything was subordinated to the immediate needs of subsistence and this, according to Bourdieu, prevented the formation of a rational temporal consciousness. As he put it in 'La Hantise du chômage ... ':

> In the absence of a regular job, what is lacking is not only a place to work and something to do every day, it is a coherent organisation of the present and the future, it is a system of expectations and a field of concrete goals in reference to which all activity can be orientated. It is only on the basis of a field of the present, which is both structured and mastered, that a future at once distant and

accessible can be aimed at and posited in a project or rational forecasting. (1962d, p. 327)

In short, Bourdieu's argument was that although they had abandoned their traditional way of life, neither the sub-proletariat nor the displaced peasants had been fully integrated into an urban, capitalist economy in which calculations of present sacrifice for possible future gain were regularly made. They remained locked within an essentially peasant experience of time, in which the future was not perceived as an immense, open field of possibilities to be realised by rational calculation, but rather was determined by the collective and cyclical rhythms of a peasant economy. It was this peasant attitude to time that Bourdieu analysed in the article 'La Société traditionnelle: attitude à l'égard du temps et conduite économique'.

Here Bourdieu distinguished between two opposing conceptions of temporality, one of which corresponded to a simple peasant economy, the other to a developed capitalist society. He employed the neologism *'l'à-venir'*, meaning 'what is to come' or 'the forthcoming', distinguishing this term both from the conventional French term for 'the future', *'l'avenir'*, and from the more specialised term *'le futur'*, which emphasises the indeterminacy of future events. In a 'traditional' peasant society, he argued, the future is grasped solely as *'un à-venir'*, as a series of future events whose possibility, attested to by past collective experience, is inscribed into the present as a 'horizon' or 'field' of 'practical possibilities' or 'objective potentialities'. This conception of the future was qualitatively different from that required to form a revolutionary 'project', a vision of a possible future which, according to Bourdieu, implied the ability to 'suspend' one's investment in the immediate self-evidence of the everyday in order to make a rational calculation of possible future gain.

Bourdieu's argument relied on a series of oppositions between the peasant's realm of temporal and perceptual *immediacy* and the possibility of reflexive *mediation*, between the realm of *self-presence* inhabited by the Algerian peasantry and the capacity to construct a mediated *re-presentation* of the social world. Each of these oppositions was then read as characteristic of the difference between a pre-capitalist and a capitalist economy. In a capitalist economy, he maintained, money was itself a medium of re-presentation, a sign or symbol which stood in for its absent referent and was thus implicated in a structure of mediation and deferral. In capitalism:

> one no longer bases one's reasoning on objects which announce, in an almost tangible and palpable manner, their function and the satisfaction they promise, but on signs which aren't in themselves the source of any satisfaction. The veil of money slips

between the economic subject and the merchandise or services he expects. (1963b, p. 33)

In a pre-capitalist economy, on the other hand, Bourdieu argued, no such medium of re-presentation existed; the peasants' perception of the social world was thus somehow immediate since the relation between objects and their function was itself immediate or self-present:

> Thus it is that *wheat gives itself immediately*, not only with its colour and its form, but also with the qualities inscribed in it as potentialities, such as 'made to be eaten'. These potentialities are grasped by a perceptual consciousness in the same way as are its directly perceived physical aspects; hence in the mode of belief [*sur le mode de la croyance*]. Whilst consciousness of a project, a consciousness involving the imagination, supposes the suspension of adherence to the given world [*une mise en suspens de l'adhésion au donné*] and aims at the projected possibilities as equally likely to be realised or not realised, the consciousness which grasps these potentialities as 'what is to come' [*à venir*] is caught up in a universe riddled with solicitations and urgent demands, the world of perception itself. The 'forthcoming' [*l'à venir*] is the concrete horizon of the present and, as such, it gives itself *in the mode of presentation not representation*, in opposition to the impersonal future [*le futur*], the site of the abstract and undetermined possibles of an interchangeable subject. (1963b, p. 37, my emphasis)

Thus, Bourdieu argued that the Algerian peasants, inhabiting a pre-capitalist social universe, perceived the world in a purely 'doxic' manner, '*sur le mode de la croyance*', as a series of immediate needs and urgent demands to be met in accordance with conventions and traditions so self-evident as to be beyond question. Any plan or project for the future would presuppose a break in this primary level of doxic adherence, in 'the passive acquiescence and spontaneous submission to the current natural and social order', as well as the accession to a notion of human agency alien to those not immersed in a capitalist mode of production:

> Far from considering himself a factor acting from outside on an external nature, man feels completely enclosed within nature. As a result, the ambition to transform the world through labour is excluded, which supposes suspending one's adherence to the natural given world [*la mise en suspens de l'adhésion au donné naturel*] and reference to an imagined and hoped for order. (1963b, p. 40)

According to Bourdieu, the Algerian peasantry inhabited a sphere of temporal immediacy in which objectives and goals could be perceived in an unmediated fashion but could not be the subject of a reflexive consciousness making rational assessments as to their potential chances of realisation. Constrained by a host of immediate, pressing needs, they were incapable of 're-presenting' a better future to themselves, of making a rational calculation of present sacrifice against possible future gain. Hence he concluded that Fanon was quite wrong to imagine the Algerian peasantry and sub-proletariat constituted revolutionary forces, emphasising in the conclusion of *Travail et travailleurs* ... that it was the 'working-class elite' which represented the only truly revolutionary class in Algeria (1963a, p. 381).

As his choice of vocabulary made clear, Bourdieu was drawing heavily on Weber's notion of the centrality of rational calculation to the ethos of Western capitalism. Indeed, he seemed to be echoing Weber's argument in *The Protestant Ethic and the Spirit of Capitalism* (1904–5) that, 'outside the modern Occident ... , the proletariat as a class could not exist, because there was no rational organisation of free labour under regular discipline ... ; because the world has known no rational organisation of labour outside the modern Occident, it has known no rational socialism' (Weber 1904–5, p. 23). Large sections of both *Travail et travailleurs* ... and *Le Déracinement* were devoted to more detailed assessments and empirical tests of the extent of the Algerian peasantry and sub-proletariat's immersion into a capitalist ethos and hence their capacity for rational calculation.

Bourdieu's analysis also owed much to a classically Marxist-Leninist theory of revolution, recalling Marx's comparison of the peasantry to 'potatoes in a sack' who could not 'represent themselves' but had to 'be represented' in *The Eighteenth Brumaire of Louis Napoleon* (1852), and Lenin's emphasis on the historical role of the proletariat in *The State and Revolution* (1918). Moreover, Bourdieu's recourse to detailed empirical research in *Travail et travailleurs* ... as a means to assess the economic and political development of Algerian society and propose a typology of its different classes and class fractions recalled Lenin's immense empirical study of *The Development of Capitalism in Russia* (1899). Bourdieu has acknowledged he was reading Marx's writings on the peasantry and 'Lenin's survey of Russia' at this time (1987, p. 17 [p. 7]).

However, Bourdieu's choice of vocabulary revealed he was also drawing on a rather different philosophical tradition and it was from this tradition that some of his more debatable assumptions about the unmediated nature of peasant experience derived. The term '*l'à-venir*', for example, is the standard French translation of the neologism '*das Kommende*' which Edmund Husserl coined in distinction to the conventional German term for 'the future', '*die Zukunft*'. Similarly, terms such as 'suspension' ['*la mise en suspens*'], 'the concrete horizon', 'the field of the present', 'the mode of belief' ['*le mode de la croyance*'], 'objective potentialities', are standard translations of the terminology employed by Husserl in his phenomenological studies of time. Bourdieu has described his essays on the Algerian peasantry and sub-proletariat as, 'researches into "the phenomenology of emotional life", or more exactly into the temporal structures of emotional experience' (1987, p. 16 [pp. 6–7]).

Husserl argues that time is not experienced as so many discrete moments in a linear causal series, but rather as a 'field' or 'network of intentionalities', of past experiences, 'retentions', which are incorporated into a structure of 'intuitive expectations' or 'practical anticipations' as so many 'protentions', a 'practical sense' of what does or does not constitute an 'objective potentiality'. He terms this structure of dispositions a 'habitus'. Habitually, in an action such as taking a spoon and placing it in the mouth, agents will have only a 'practical' sense of what they are doing; they will intuitively anticipate the position of their own mouth, for example, based on practical aptitudes they have picked up through past experience. Such a sense requires no reflection on their part on the complex range of physiological and mental capacities involved (Husserl 1952, pp. 266–93). Husserl thus distinguishes between this practical sense of what is to come, 'the intuitively expected thing, of which, by anticipation, we are conscious as being what is to come [*comme "à venir"*]' and the 'modification', 'suspension', or '*epoche*' in this 'natural' or 'doxic' attitude which allows for a more reflexive grasp of the objective processes at work in any action (1913, pp. 242–99).

Bourdieu's argument in relation to the Algerian sub-proletariat and peasantry was that their material state, their need to satisfy the immediate demands of subsistence, prevented them from moving beyond a purely practical or doxic apprehension of their social universe to gain a more reflexive awareness either of the logic of their current behaviour or of possible alternative modes of behaviour. The enforced migrations of the sub-proletariat and peasantry may have provoked a suspension in their immediate adherence to old conventions and traditions, but their continuing economic insecurity and sense of cultural dislocation prevented them from integrating completely into a rational, capitalist mode of social and cultural organisation, the precondition, according to Bourdieu, for effecting that transition from the doxic to the reflexive mode and hence gaining a truly revolutionary consciousness.

Within this account of the internalised expectations and field of objective possibilities open to the Algerian masses was contained the first tentative sketch of what would later become Bourdieu's theory of 'habitus', 'practice' and 'field'. According to this theory of social action, agents are neither totally free nor the mere puppets of objective social laws. Rather, they 'incorporate' a 'practical sense' of what can or cannot be achieved, based on intuitions gained through past collective experience, into their 'habitus', a structure of dispositions which thus reflects the 'field of objective possibilities' open to them at a particular historical moment. As Bourdieu has put it, behind the concept of 'habitus' lies the principle,

> of a knowledge without consciousness, of an intentionality without intention, and of a practical mastery of the world's regularities which allows one to anticipate the future without even needing to posit it as such. We find here the foundations of the difference established by Husserl in *Ideen I*, between the protention as the practical aiming at what is to come [*un à-venir*] inscribed in the present, thus apprehended as already there and endowed with the doxic modality of the present, and the project as the positing of a future [*un futur*] constituted as such, that is capable of happening or not happening. (1987, p. 22 [p. 12, trans. modified])

This early formulation of the theory of habitus and practice also involved Bourdieu in a polemic with Sartre's theory of revolutionary action. In his *The Problem of Method*, Sartre had also drawn on Husserlian phenomenology, grafting it on to a Hegelian reading of Marx, to elaborate what he termed 'the progressive-regressive method'. Drawing on Husserl's concepts of retention and protention, Sartre argued that human agents were the products of a history which they internalised and sublated (preserved, negated, and transcended) in their future actions or 'projects'. In Sartre's rather teleological and subjectivist account, oppressed individuals, because of their feelings of alienation, would engage in revolutionary praxis, in a 'project' which aimed to overcome that alienation:

> to feel is already to transcend, to move towards the possibility of an objective transformation. *Faced with the trials of lived experience*, subjectivity turns back on itself and wrenches itself from despair by means of *objectification*. Thus the subjective preserves within itself the objective, which it negates and which it transcends towards a new objectivity; and this new objectivity, by virtue of *objectification*, externalises the internality of the project as an objectified subjectivity. (1960, pp. 97–8 [trans. modified])

Throughout *The Problem of Method*, Sartre invoked the example of colonised peoples who, denied rights and opportunities by the colonial system, would fight to overcome such obstacles by engaging in revolutionary praxis, a struggle for independence that was simultaneously a revolutionary transcendence of their alienated state.

In 'Les Sous-prolétaires algériens', Bourdieu argued, against Sartre, that the Algerian sub-proletariat were alienated to such an extent as to be unable to gain objective consciousness of their subjective state: 'absolute alienation deprives the individual of even the consciousness of alienation' (1962e, p. 1049). Thus they were unable to transcend their alienation through revolutionary praxis. (In the quotation below, the phrase 'a sort of revolutionary *cogito*' is typical Bourdieusian shorthand for Sartre's theory of revolutionary action).

> to constitute the current state of affairs as possessing this or that meaning supposes something quite different from a sort of revolutionary *cogito* by which consciousness would make itself revolutionary by wrenching itself from the world, a world to which consciousness is present but which it cannot represent because it is enclosed within the world and transcended by the world. If it is true that to constitute the current state of affairs as intolerable and revolting supposes positing another state of affairs, at once absent and accessible, it remains true that even to posit a possible alternative presupposes the possibility of taking a certain distance from the world. (1962e, p. 1051)

It was precisely this reflexive distance which was unavailable to the sub-proletariat, Bourdieu maintained: 'In short, reflection demands a certain material well-being and, paradoxically, political awareness [*la prise de conscience*] is a privilege incumbent on those who are not so alienated that they cannot stand back and take their situation in hand' (p. 1051). As he argued in *Le Déracinement*, it was only by ignoring the material conditions of the peasantry and sub-proletariat that one could lend 'any credence to the eschatological prophesies which see in the peasantry of the colonised countries the only truly revolutionary class' (1964, p. 170).

The desire to avoid an eschatological account of the political development of the Algerian masses or indeed a teleological account of the modernisation of Algerian society was at the heart of both *Travail et travailleurs ...* and *Le Déracinement. Travail et travailleurs ...* was divided into two separate but interrelated books. In the first book, Bourdieu's co-authors presented the results of a statistical survey which defined the principal characteristics, in terms of age, sex, profession, socio-economic category, and so on, of the populations of Algeria's shantytowns and 'resettlement centres'. In the second, Bourdieu offered a 'sociological survey' of the attitudes of those populations to work and unemployment, which drew both on data culled from questionnaires and on more ethnographic fieldwork.

Where the statistical sections of *Travail et travailleurs ...* offered an objective measure of the modernisation and rationalisation of

Algerian society, the more sociological sections written by Bourdieu sought to examine how such apparently impersonal processes of modernisation were mediated through the subjective dispositions of the Algerian masses themselves. These dispositions had, for the most part, been formed under an earlier model of 'traditional' society whose characteristics could be identified by anthropological analysis. Anthropological study of 'traditional' Algerian society was thus to combine with statistical surveys and ethnographic research into contemporary Algeria to allow for an assessment of the complex dialectic of tradition and modernity. Attitudes to work and unemployment were seen to be at the centre of this dialectic since, Bourdieu argued, awareness of unemployment was an objective measure of the Algerians' 'break' with a 'traditional' society in which there was no sense of the division between productive and non-productive labour:

> in the colonial situation, work is the site *par excellence* of the conflict between traditional models and the models imported and imposed by colonisation, or, if you prefer, between the imperatives of ration-alisation and cultural traditions. It follows that understanding statistical regularities is only possible provided that, thanks to anthropology, we can reconstruct the model from which current behaviour derives its meaning, even at the moment it betrays or repudiates that model. Every act is related at the very same moment to the old model, which formed part of a system now partially or totally destroyed, to the new situation, and finally to the model to come, which can be glimpsed through the bizarre or contradictory nature of present behaviour. (1963a, p. 266)

Bourdieu's account of the relationship between statistical analysis, sociological survey and anthropological research was an early example of his attempt to overcome the opposition between 'objectivist' and 'subjectivist' accounts of the social world. As has already been seen, this attempt involved sketching out what would later become the key Bourdieusian concept of 'habitus', a concept which sought to describe the way objective or material conditions of existence were internalised into a structure of subjective dispositions, a set of practical expectations or anticipations, an 'attitude towards time' which reflected the 'objective future', the 'field of effective possibilities' open to particular agents or groups (1963a, p. 346).

By focusing on the way in which the objective processes of modernisation were mediated through subjective dispositions, which themselves reflected a specific relationship to 'traditional' social and cultural forms, Bourdieu was able to offer a much more nuanced account of the relationship between tradition and modernity than he had suggested in *Sociologie de l'Algérie*. By distinguishing between the young, who had relatively little attachment to 'traditional' forms, and the old, who were deeply attached to them, between the more and the less educated groups, between those, such as the Kabyles, who had a history of labour migration to France, and those who did not, Bourdieu offered a highly variegated account of Algeria's political and economic development.

This more nuanced account of modernisation in Algeria was also evident in *Le Déracinement*. Co-written with Abdelmalek Sayad, *Le Déracinement* was a detailed empirical study of the effects of the French authorities' policy of 'resettlement' on peasants from different regions of Algeria. The text can be read as a narrative of Weberian 'disenchantment' in which Bourdieu described peasants uprooted from their native villages and forcibly relocated into 'resettlement centres' where older customs and traditions were eroded by a rational, regimented organisation of time, space and economy. Bourdieu seemed here to pay more attention to Weber's warnings of 'the danger that ideal-type and reality will be confused with one another', of the need to see the ideal-typical 'reconstruction' of 'traditional' societies as a purely heuristic device which corresponded to no actually existing society but was to be used merely for 'the comparison of the ideal-type and the "facts"' (Weber 1949, pp. 101–2).

Thus Bourdieu emphasised that 'the opposition between the traditionalist peasant and the modern peasant now has only a heuristic value and merely defines the extreme poles of a *continuum* of behaviour and attitudes separated by an infinity of infinitesimal differences' (1964, p. 161). By paying close attention to the different histories of the various groups of peasants he studied, the extent of their contacts with colonial society, their history of emigration to France, their education, and their differing levels of political and social development, Bourdieu was able to analyse that 'infinity of infinitesimal differences'. 'Traditionalism', previously seen as the 'internal logic' of Algerian tribal society, was now described as an 'enforced traditionalism', a 'pathological' response to colonialism. Colonialism had deprived the Algerians of the possibility open to other societies to choose elements of a 'modern' or 'rationalised' society whilst adapting them in accordance with indigenous traditions and customs (p. 34).

Bourdieu's account of the dialectical relationship between 'tradition' and 'modernity' in *Travail et travailleurs* ... and *Le Déracinement* suggested he was moving beyond the model of a unidirectional process of cultural change implicit in classical theories of 'acculturation'. Indeed, in the Introduction he wrote to the excerpts from *Travail et travailleurs* ... included in *Algeria 1960*, Bourdieu criticised acculturation theories for their abstraction and inability to grasp the complex dialectic of tradition and modernity at work in any process of rationalisation and modernisation (pp. 1–2). Although he never used the term, it might be argued that Bourdieu was working towards a theory of 'transculturation'.

However, Bourdieu's interest in Algerian modernisation was not purely theoretical. On the contrary, his analyses were motivated by urgent political questions. As he pointed out in a footnote (1963a, p. 380n.1), the final sections of *Travail et travailleurs* ... were written in May 1963, just two months after the March Decrees which heralded the first major wave of nationalisations of Algeria's major industries and of large tracts of agricultural land. Industry and agriculture alike were to be run by a system of 'autogestion' or workers' self-management, a policy which could find justification in Fanon's theories regarding the revolutionary role of the peasantry and sub-proletariat. In *Travail et travailleurs* ... , Bourdieu was attempting to show how ill-equipped both peasantry and sub-proletariat were to fulfil any such role. The lack of a developed political consciousness amongst the peasantry and the sub-proletariat would leave a political void, which, he argued, would inevitably be filled by a semi-educated petty bourgeoisie and a bourgeois intellectual elite both of whom had a vested interested in strengthening the power of the centralised state bureaucracy in which they typically worked. By the time *Le Déracinement* was published, Bourdieu's predictions seemed to be coming true and he noted that the inability of the peasantry and sub-proletariat to fulfil the unrealistic role allotted them had provoked a slide away from '*autogestion*' towards centralised bureaucracy and 'an authoritarian socialism, the army's preferred solution' (1964, p. 169). It was in the face of this authoritarian turn that in *Travail et travailleurs* ... he had emphasised, in classically Leninist style, that it was only the working class, 'the workers' elite', which expressed 'rational and *universal* demands and claims' (1963a, p. 381). In *Le Déracinement*, he called for 'a complete and total action of education', to ensure the Algerian peasantry's transition from tradition to modernity, overseen by 'a revolutionary elite' sensitive to the peasantry's current state and real needs (1964, pp. 176–7).

With the benefit of hindsight, it is possible to see that Bourdieu's concerns about the March Decrees were fully justified; numerous commentators have interpreted them as a key moment in Algeria's descent into centralised bureaucratic sclerosis.[5] However, it should be noted that Bourdieu was not alone in correctly anticipating the outcome of the Algerian Revolution. As early as 1956, Jean-François Lyotard had warned of the dangers of post-independence Algeria, 'in the absence of any proletarian consciousness', descending into 'an embryonic military and political bureaucracy to which the diverse elements of the Moslem intellectual and commercial stratum will be likely to rally'. This was a theme Lyotard developed, with remarkable prescience, in a series of articles published in the journal *Socialisme ou barbarie* between 1956 and 1963.[6] Furthermore, Fanon himself had anticipated the possibility of post-colonial governments becoming centralised state bureaucracies dominated by the educated urban petty bourgeoisie and bourgeoisie, dedicating an entire chapter to this problem in *The Wretched of the Earth* (Fanon 1961, pp. 119–65). As Jean-Marie Domenach (1962) has pointed out, Fanon's emphasis on the role of the peasantry and sub-proletariat was intended precisely as a precaution against the possible hegemony of a self-interested urban indigenous bourgeoisie concerned only with the pursuit of industrial development.

Bourdieu was clearly justified in his critique of the dangers implicit in Fanon's portrayal of peasantry and sub-proletariat as revolutionary classes. However, in using the capacity for rational *economic* calculation as a straightforward measure of the Algerian masses' capacity for rational action in the *political* domain, he risked implying that the key to the Algerians' further political development lay primarily in economic development and industrial modernisation.[7] Bourdieu has always insisted that his work struck a blow against such economism by refusing to assume that the actions of all agents in every historical situation could be judged according to the ahistorical norm of the *homo economicus*, of an agent who makes decisions based on rational assessments of future profit or loss. However, Bourdieu's argument rested on the premise that there was a direct and unproblematic relationship between the level of the Algerians' *economic* development, their capacity for rational economic calculations, and their capacity for rational action in the *political* domain. As Bourdieu put it, the adoption of 'the spirit of forecasting and the ability to construct a rational [political] project':

> varies in direct proportion to *the degree of integration* within an economic and social order defined by the capacity for calculation and forecasting ... . Because the *objective future* of individuals and groups depends very closely on their material conditions of existence, it is obvious that the free project and calculated forecasting are privileges which are inseparable from a certain economic and social condition; internalisation of the objective future, a plan for life is reserved to those whose present and future have already been wrenched from incoherence. (1963a, pp. 366–7)

The conclusion to be drawn from such passages seemed to be that the more economically advanced an agent or class was, the greater its capacity for rational political agency. What was lacking from this account was any distinction between the economic and political domains, any acknowledgement of the complex mediations between the economic and the political.

It was precisely this suggestion of an unmediated link between economic development and political agency that Nicos Poulantzas (1968, pp. 60–5) criticised in his remarks on *Travail et travailleurs … in Political Power and Social Classes* (*Pouvoir politique et classes sociales*, vol.1 1968). According to Poulantzas, Bourdieu's argument was a classic example of the link between economism and humanism which Louis Althusser and his co-authors had identified in their 'structural' readings of Marx. For the Althusserians, it was a humanist and economistic reading of Marx which had led to the bureaucratic sclerosis and over-emphasis on increased industrial production which characterised Stalinism. They sought instead to emphasise the ways in which economic change was mediated through and articulated with political and other superstructural 'instances', whilst abandoning an evolutionary model of history based on a straightforward opposition between tradition and modernity in favour of a vision of different and differential historical temporalities (Althusser 1965 and 1965a).

In the Introduction to *Algeria 1960*, Bourdieu would argue that his account of Algeria's transition from tradition to modernity was preferable to that offered by structural Marxism, arguing, with some justification, that the latter had remained excessively abstract (p. vii). Nonetheless, his own analysis of economic and political development in Algeria might have benefited from a greater emphasis on the mediations between the economic and political domains and hence a greater reluctance to link economic development and political consciousness quite so directly. Ultimately, though, Bourdieu's argument relied both on maintaining that link and on mobilising a series of oppositions between the 'immediacy' of the Algerian peasantry's and sub-proletariat's temporal and political consciousness, their inability to reflect on their lives or construct a rational representation of a possible alternative future, and the capacity for reflexive mediation and rational forecasting open to those who had been integrated into a 'modern' Western capitalist economy. There was surely a danger here of reproducing a set of profoundly ethnocentric oppositions between the 'modern', 'rational', dynamic West and the 'primitive', 'pre-rational', immobile non-West. Certainly such ethnocentrism was implicit in Husserl's phenomenological studies of time, to which Bourdieu's analyses were so greatly indebted.

Husserl's distinctions between 'the forthcoming' (*l'à venir*) and 'the future', between 'doxic' or 'practical' knowledge and theoretical knowledge were doubled throughout his work by an analogous opposition between the non-West and the West. Husserl believed that there was a congenital difference between the 'temporal consciousness' of 'primitive', non-Western peoples and 'rational' Western peoples.[8] As Paul de Man (1971, pp. 15–16) puts it:

'Husserl speaks repeatedly of non-European cultures as primitive, pre-scientific and pre-philosophical, myth-dominated and congenitally incapable of the disinterested distance without which there can be no philosophical mediation ... , he warns us, with the noblest of intentions, that we should not assume a potential for philosophical attitudes in non-European cultures.' It could be argued that this was precisely Bourdieu's argument in relation to Sartre and Fanon, namely that they were wrong to assume a potential for philosophical mediation, disinterested distance or political reflection amongst the Algerian masses. This line of argument would re-emerge in a notoriously controversial passage in *Distinction* in which Bourdieu argued that it was wrong to attribute the French working class with any genuine culture of their own since the immediacy of their material need rendered them incapable of the disinterested distance necessary for aesthetic contemplation (1979, pp. 433–61 [pp. 372–96]).

The very fact that Bourdieu attributed the same lack of reflexive distance to the French working class as to the Algerian masses indicates that he was not simply reproducing Husserl's ethnocentric belief in a *congenital* difference between the 'primitive' and the 'modern' mentalities. As Bourdieu himself highlighted, it was vital to 'be on guard against ethnocentrism and against the inclination to describe this temporal consciousness of pre-capitalist man as being separate by nature from that which the capitalist man forms. In fact temporal consciousness depends on the ethos specific to each civilisation' (1963b, p. 38). However, as has been demonstrated, the manner in which Bourdieu theorised the relationship between temporal consciousness and material circumstances would lead him into a form of economism.

This residual economism must ultimately be attributed to Bourdieu's reliance, from *Sociologie de l'Algérie* to *Travail et travailleurs* ... and *Le Déracinement*, on a series of rather stark dichotomies between tradition and modernity, immediacy and mediation, presence and representation. To point to the persistence of such dichotomies is by no means to seek to write off all of his work of the period as simply 'wrong' or a 'failure'. On the contrary, it is merely to suggest that the line of thought initiated by Bourdieu in works such as *Travail et travailleurs ...*, the attempt to think beyond the opposition between objectivism and subjectivism, the emphasis on grasping processes of change and modernisation in all their dialectical complexity, and so on, could be pursued still further in ways that might force a re-formulation of his terms of analysis, of the problematic within which he was working.

Having returned to France in the early 1960s, Bourdieu was to be caught up in a series of debates concerning the precise significance of the social and cultural changes wrought by France's postwar economic prosperity. In his contributions to these debates, he would draw on and extend the theoretical approaches he had first sketched in Algeria, refining the concept of 'habitus', for example, to analyse the ways in which modernisation in postwar France was being mediated. In his early work on French culture and society, Bourdieu was to adopt far more reformist political positions than he had in Algeria. Indeed, his early works on Algeria have proved the only occasion on which he has explicitly advocated revolutionary solutions to the social phenomena he studies.

Nonetheless, Bourdieu's contributions to debates on postwar French modernisation did reveal a series of continuities with the approaches he had pioneered in his work on Algeria, which will be discussed in the next chapter.

Notes to Chapter 1

1. Although Bourdieu was to dedicate *The Algerians* (1962) to Aron, their friendship and collaboration should not be taken to indicate any particular *political* affinity, either as regards the Algerian War or any other contemporary issue. Generally speaking, Aron's approach to the Algerian question was both less radical and more pragmatic than Bourdieu's (see Aron 1957 and 1958).
2. For a description of the 'objectivity' of sociological enquiry and the role of 'ideal-types' in 'historical reconstruction', see Weber (1949, pp. 50–112).
3. Bourdieu was drawing on the work of American acculturation theorists Melville Herskovits, Ralph Linton, and Robert Redfield (see Herskovits 1938).
4. See also Bennoune (1988, pp. 15–31) and Lacoste-Dujardin (1997).
5. For an analysis of the March Decrees and their aftermath, see Chaliand and Minces (1972, pp. 32–59); Jackson (1977, pp. 135–54); and Tlemcani (1986, pp. 95–107).
6. These articles have been anthologised in Lyotard (1989).
7. As Bourdieu himself has indicated, *Travail et travailleurs* ... and *Le Déracinement* were both invoked, against his express intentions, in support of the rapid drive for industrialisation pursued by the post-independence Algerian regime (1980, p. 8 [p. 2]).
8. See the excerpts from Husserl's letter to Lucien Lévy-Bruhl quoted by Merleau-Ponty (1960, p. 135).

CHAPTER 2

Frenchmen into Consumers?

If Bourdieu's experiences during the Algerian War had led him to undertake what he had initially considered to be a temporary detour into sociology, by the mid-1960s he had become clearly established as a professional sociologist occupying a position of considerable institutional prestige, surrounded by a team of collaborators. After a year spent as Raymond Aron's assistant at the Sorbonne, Bourdieu had been appointed as a sociology lecturer at the University of Lille in 1961, working there until 1964. During this period, he became secretary of the Centre de sociologie européenne (CSE), a loose grouping of researchers established by Aron. In 1964, he became Directeur d'études at the École pratique des hautes études en sciences sociales in Paris, a post he has continued to occupy even after his appointment as professor at the Collège de France in 1981.

Bourdieu's new status was reflected in a broadening of his research interests and the 1960s saw him publish a series of articles and book-length studies of contemporary developments in French culture and society. This new body of work grew out of various research seminars at the CSE and the books themselves were written in collaboration with other researchers attached to the Centre. As Bourdieu has remarked, the studies of education, class, and culture he published under the aegis of the CSE were 'born of a generalisation of the results of the ethnological and sociological studies I had carried out in Algeria' (1987, p. 33 [p. 23]). Bourdieu's work on education will be discussed in the next chapter. This chapter will focus on his early analyses of the relationship between class and culture, examining how the approach he had sketched in his Algerian work was applied and refined in this new context. This work on class and culture ranged from his studies of photography and the art gallery, *Photography: a middle-brow art* (*Un art moyen* 1965) and *The Love of Art* (*L'Amour de l'art* 1966), to less well-known analyses of the effects of France's postwar economic boom, whether on the peasant communities of his native Béarn, in the long essay 'Célibat et condition paysanne' (Celibacy and the Peasant Condition) (1962), or on French society as a whole, in his contributions to the collection, *Le Partage des bénéfices* (Sharing the Spoils) (1966).

These works were published at the height of 'les trente glorieuses', that thirty-year period of demographic renewal, reconstruction, modernisation, and economic growth that marked France's emergence from the Second World War. From 1959 to 1970, France's GDP achieved an average growth of 5.8 per cent, second only to Japan amongst the Group Seven nations. This growth was accompanied by an increase in disposable incomes (from 1958 to 1969 average incomes rose by 50 per cent in real terms) and by a reduction in working hours and an increase in paid holidays. This, combined with the new consumer credit facilities, meant that French consumers had both more money to spend on consumer durables such as cars, radios, fridges and washing machines and more time available for leisure and holidays. Between 1958 and 1969, household expenditure on leisure and cultural activities increased by an average of 50 per cent.

This period of economic reconstruction and growth, which coincided with French decolonisation, can be read as symptomatic of that shift from 'imperialist' to 'late capitalism' which Ernest Mandel (1975) argues affected the major Western economies in the postwar period. Decolonisation was accompanied by a rapid decline in trade between France and its colonies or former colonies and its replacement by an emphasis on the sale of an ever-increasing range of consumer goods within France itself or between France and other developed economies. This demanded the rationalisation and modernisation of French industry's production and distribution processes, whilst heralding the growth of a large tertiary sector and the emergence of new strata of middle managers, advertisers, research and development executives and market researchers, who were charged with locating new markets and manipulating consumer demand. It was these new strata who formed the core of the aspirational and upwardly mobile '*classes moyennes*' or lower middle classes, the subject of so much interest in the journalistic, novelistic, and sociological discourses of the day (Ross 1995, pp. 126–45).

As Mandel (1975, pp. 500–22) argues, the advent of late capitalism is typically accompanied at the ideological level by a proliferation of discourses proclaiming the ability of technological progress and planned economic growth to end class conflict and ensure the progressive satisfaction of all material and spiritual needs. Whilst Bourdieu never theorised this shift to late capitalism as such, this chapter will argue that his work of the period was centrally concerned with challenging such technocratic discourses by attempting to achieve a more objective measure of the extent or limitations of the social and cultural changes wrought by France's postwar economic growth. His early works on class and culture can perhaps be best understood as interventions into contemporary debates surrounding the extent to which France's postwar economic expansion was heralding the advent of a mass consumer culture, an era of cultural 'democratisation' and 'homogenisation' in which older class divisions were either disappearing or being significantly redrawn.

Having sketched the sociological field into which Bourdieu's work intervened and the competing discourses against which his position defined itself, this chapter will analyse the nature of his challenge to those discourses. It will show how this challenge often mirrored approaches Bourdieu had outlined in his work on Algeria, involving a return to the classical sociological tradition of Marx, Weber, and, most significantly, Durkheim and the Durkheimians, as well as an emphasis on the importance of empirical research.

Technocrats and 'Massmediologists'

In an article of 1967 he co-authored with Jean-Claude Passeron, entitled 'Sociology and Philosophy in France since 1945: Death and Resurrection of a Philosophy without Subject', Bourdieu surveyed postwar developments in the French intellectual field. The principal development he identified was the rise of the French social sciences, manifest in the rapidly increasing number of sociological researchers and research bodies, the appearance of a raft of new social science journals, and the recognition of the social sciences as autonomous disciplines in French universities for the first time (1967a, pp. 180–94). Bourdieu did not, however, welcome such developments unreservedly, noting that the boom in the social sciences owed much to 'the rather large "orders" in the field of sociological research' placed by France's state planning agency, the Commissariat au Plan, often known simply as 'the Plan' (p. 186).

With the encouragement of state planners and against the background of 'the economic prosperity and concomitant security' of the 1950s and 1960s, the discipline of sociology was being inflected away from a concern with problems of deprivation, poverty, or class towards an emphasis on the 'autonomous and anonymous efficiency' of economic systems and business organisations, on systematicity and synchronicity over 'history and the subject as agent of history'. This shift was manifest in the vogue for structuralism, 'the development of an empirical sociology of a positivist kind', or the publication of books such as Michel Crozier's *The Bureaucratic Phenomenon* (*Le Phénomène bureaucratique* 1963), itself funded by a grant from the Commissariat général à la productivité (pp. 191–2). Bourdieu was concerned that the social sciences were being subordinated to the technocratic demands of state planning. He noted with dismay the participation of high-ranking technocrats such as Claude Gruson and Pierre Massé at the 1965 Congrès de la société française de la sociologie, as well as the proliferation of 'new exchange places ... where senior officials and sociologists of the upper echelon of the administration' met to 'pool their ideas' (1967a, p. 187).

It is surely the writings of Jean Fourastié which best exemplify the technocratic faith in the ability of planned economic growth, informed by the social sciences, to resolve all of France's social and political problems. President since 1948 of the working group on productivity at the Plan, Fourastié published a series of popular sociological studies in the postwar period, notably *Le Grand Espoir du XXe Siècle* (The Great Hope of the 20th Century), originally published in 1949 and reprinted in several revised editions until the 'definitive edition' of 1963. He used France's economic performance, the widening access to education, the improvements in housing conditions and standards of living, as the basis upon which to construct a utopian vision of the meritocratic society just around the corner, a 'society without classes' on the American model in which modernisation and economic growth would signal an end to social conflict.

Bourdieu's criticisms of this kind of technocratic sociology might seem to place him in alliance with a range of contemporary thinkers on the French Left, sociologists such as Henri Lefebvre, Alain Touraine, Edgar Morin, and Roland Barthes, who were equally critical of technocracy. Indeed, in his *Position: contre les technocrates* (Position: against the technocrats) (1967), Lefebvre pointed to the same affinity between structuralism and technocracy as Bourdieu noted in his 'Sociology and Philosophy ... ' of the same year. However, far from signalling his proximity to such thinkers, Bourdieu argued that they, and the political programmes of the French 'New' or 'Second Left' with which they were associated, shared with the technocrats an exaggerated faith in the ability of planned economic growth and technological modernisation to provoke significant social change. In *Reproduction*, Bourdieu would write dismissively of certain '"critical" ideologies', which, in their fascination with 'the homogenising and alienating effects of television or the "mass media"', shared the same blindness to questions of class, 'an indifference to differences', as 'their favourite enemy, technocracy' (1970, p. 229n.15 [p. 213n.15]). Without mentioning anyone by name, he lambasted a series of thinkers; 'the shock troops of every avant-garde, who ceaselessly scan the horizon of "modernity" for fear of missing an ideological or theoretical revolution, ever ready to spot the newest born of the "new classes", "new alienations" or "new contradictions"' (Bourdieu 1970, p. 228 [p. 193]).

Although Bourdieu refrained from naming the targets of his critique, his choice of vocabulary made it clear whom he had in mind. For it was Lefebvre who had first diagnosed the alienating and reifying effects of 'mass culture' in his various postwar studies of everyday life and modernity. These studies exerted a considerable influence over French sociology of the late 1950s and 1960s, as manifest in Touraine's attempts to enumerate the different forms of 'alienation' characteristic of modernity in his *La Sociologie de l'action* (Sociology of Action) (1965), as well as in Barthes's semiological readings of consumer culture in his massively influential *Mythologies* (1957). Morin's study, *L'Esprit du temps* (The Spirit of the Times) (1962), is less well-known outside France but its analyses of the simultaneous danger and promise of the alienating 'mass media', at once a 'homogenising' and potentially 'democratising' force, enjoyed considerable influence at the time. Finally, Bourdieu's reference to 'the newest born of the "new classes"' was clearly an allusion to two influential studies of the emergence of a highly qualified 'new working class' in France's technologically advanced industries, Serge Mallet's *The New Working Class* (*La nouvelle classe ouvrière* 1963) and Pierre Belleville's *Une nouvelle classe ouvrière* (1963). These influential studies had fed into Touraine's *Post-industrial Society* (*La Société post-industrielle* 1969), in which he maintained that the emergence of this 'new working class' signalled an end to the old struggle between capitalist and worker over ownership of the means of production and its replacement by a struggle between technicians and technocrats over ownership of knowledge.

Bourdieu's critique of Touraine, Lefebvre, Morin, Barthes, Mallet, and Belleville turned on questions of both politics and sociological methodology. Politically, these figures were all associated with the French New Left, a loose grouping of intellectuals, politicians, and trade unionists who had rejected the Stalinism of the French Communist Party and were attempting to found a more democratic socialist movement, a 'third way' between Stalinist communism and unfettered liberal capitalism. They gravitated around journals such as *Esprit*, *Les Temps modernes*, and *Arguments*, whilst their ideas fed into the programmes of the breakaway socialist *groupuscule* the Parti socialiste unifié (PSU) and the modernising current within the non-Communist trade union the Confédération française démocratique du travail (CFDT). In the mid-1960s, democratically controlled state planning was at the centre of the political programmes of the PSU and CFDT alike. Indeed, the PSU's most famous member, Pierre Mendès France, had himself been involved in the Plan at the Liberation and was known as an arch moderniser. Similarly, the young Jacques Delors, one of the chief political theorists of the CFDT's modernising current, was employed at the Plan. Both Belleville and Mallet were leading figures in the PSU. Moreover, Mallet's study of the 'new working class' had been conducted under the aegis of the Club Jean Moulin, a group of high-ranking civil servants with one foot in the camp of Charles de Gaulle and the other in that of Mendès France, one of the 'new exchange places' between technocrats and sociologists cited by Bourdieu in 'Sociology and Philosophy ... '.[1]

The journal *Esprit* provided Bourdieu with further evidence of the affinities between technocracy and the New Left. *Esprit*'s founder, Emmanuel Mounier, believed in the need for France to be led by an elite of spiritual men and women, a belief which could all too easily translate into support for technocracy. Moreover, as Mounier had led this progressive Catholic journal to the Left in the postwar period, he had attempted to open a dialogue between Marxism and the writings of the Jesuit priest and philosopher Pierre Teilhard de Chardin, embracing a mix of corporatist economics and faith in technological progress close to the philosophy of the state planners. As Bourdieu (1967a, p. 200) pointed out, Teilhard de Chardin's 'theology of growth' was highly influential with French state planners at this time.[2]

For Bourdieu, the New Left's belief in the ability of planned economic growth and technological progress to provoke positive social change reflected the methodological failings of its favourite sociologists, particularly that 'marginal intelligentsia' associated with the journal *Arguments* (1967a, pp. 181–2). Those involved in *Arguments*, such as Lefebvre, Barthes, Morin, and Pierre Fougeyrollas, had been responsible for the first studies of the effects of 'mass consumerism' and the 'mass media' on French society (Stafford 1997). They focused on the images of consumerism and modernity transmitted through the mass media, through advertising, Hollywood cinema, radio, television, and the photo-journalism of magazines such as *Elle*, *Paris Match*, *L'Express*, and *Madame Express*. In the article 'Sociologues des mythologies et mythologies des sociologues' (Sociologists of Mythologies and Mythologies of Sociologists) (1963), again co-authored with Passeron, Bourdieu had mockingly dubbed the authors of such studies 'massmediologists', singling out Barthes's *Mythologies* and Morin's *L'Esprit du temps* for particular criticism. Rather than base their analyses on empirical research into the modes of production and reception of such images, he argued, the 'massmediologists' fell back on a kind of technological determinism in which the existence of such new 'mass media' was taken in itself as evidence of the prevalence of an alienating 'mass culture' being passively consumed by the docile 'masses' (1963c, p. 1002).

Freed from the burden of undertaking detailed empirical research into the way in which mass cultural forms were actually being produced or consumed, the 'massmediologists' could secure the symbolic profits attendant upon a hermeneutic style of structuralist or semiotic analysis. In an ostentatious and typically intellectual display of '*prophétisme*' or interpretative brilliance, they could reveal the 'truth' behind the ideological mystifications which held the everyone else in their thrall (p. 1000). Unable 'to reduce a providential pretext for prophetic prestidigitation to a vulgar object of science' (p. 1020), the 'massmediologists' refused to adopt a genuinely sociological and scientific approach of the kind that Bourdieu would claim to undertake in the works on class and culture he published in the course of the 1960s. These works would seek to correct the hasty analyses not only of technocrats such as Fourastié, but also of the 'massmediologists', of Lefebvre, Morin, Mallet, and others.

As in his early work on Algeria, so in his early studies of French society, Bourdieu was staking a claim to the pre-eminence of an empirically informed sociology over other forms of analysis which, although more intellectually prestigious, were to his mind less scientific. At the heart of both his work on Algeria and his studies of France was the question of economic modernisation and the need to grasp the structures through which that modernisation was mediated. In his arguments with the technocrats and 'massmediologists' about the extent of the social and cultural changes sweeping through postwar France, Bourdieu would place an analogous emphasis on the way in which technological change and increasing economic prosperity were mediated through and mitigated by the structures of a class determined habitus. In *Photography*, for example, he would demonstrate how a cultural practice, which technological progress had rendered affordable to all but the most impoverished in French society, nonetheless continued to be the site of class divisions and distinctions, which themselves reflected the workings of class ethos and habitus. It was, however, the long article on the changes sweeping through the peasant communities of Bourdieu's native Béarn, 'Célibat et condition paysanne' (1962), which offered the most striking parallels with his studies of Algerian society. For Bourdieu clearly saw the erosion of the peasant way of life in the Béarn as analogous to the upheavals in 'traditional' peasant society he had witnessed in Algeria.[3]

From the exode rural *to the* grands ensembles

Just as Bourdieu had found that in Algeria it was the peasants who were most acutely affected by the turmoil of the War of Independence, so in France after 1945 it was the rural population which experienced some of the most profound upheavals. *L'exode rural*, the mass migration of peasants into French towns and cities in search of work, was perhaps the most visible manifestation of such upheaval. Between 1946 and 1968, the number of French workers employed on the land fell by over four million, reducing the percentage of the working population employed in agriculture from 36 per cent to just 15 per cent. In the long article 'Célibat et condition paysanne', Bourdieu analysed the effects of this rural exodus on the peasant communities of his native Béarn. His starting point was a marked increase in the rate of enforced celibacy amongst elder sons, traditionally the most prized marriage partners since they stood to inherit their father's farm. Increasingly, those who had greatest value on the marriage market were not those who stood to inherit land but those who lived in urban areas. Modern, urban values were eroding that carefully constructed set of moral precepts and social conventions which had traditionally regulated peasant life, the peasant 'ethos' which set collective limits on ambition and future action, securing the integration of the whole group. Accumulated wealth, which would once have been invested in land, was now invested in 'the accumulation ... or the ostentation of consumer goods such as a car or a television' (Bourdieu 1962c, p. 83).

Bourdieu paid particular attention to the role of young women as the ambassadors for this set of modern urban social mores. Unlike their male counterparts, who were forced to stay at home to work the land, young women had been able to take advantage of the extension of formal education, often moving to the local town to study or work, where they were exposed to a range of other cultural influences. Through women's magazines, cinema, the radio and pop music, young women were acquiring 'models of the relations between the sexes and of an ideal man who is the complete opposite of the traditional peasant [*le paysan "empaysanné"*]' (Bourdieu 1962c, pp. 105–6). If the young Béarnaise women now saw their male counterparts as hopelessly clumsy and unattractive, the men of the Béarn had themselves internalised such unflattering perceptions into both their awkward bearing and manner, their 'bodily hexis', and their low expectations of ever finding a marriage partner, 'their habitus'. The male peasants had adjusted their expectations in line with their declining objective chances of finding a marriage partner so as effectively to exclude themselves from the marriage market altogether. This was Bourdieu's first use of the terms 'hexis' and 'habitus', which have taken on an increasing importance in his later work. The Greek term '*hexis*' is originally derived from Aristotle and means simply 'habit'. '*Habitus*' is its Latin equivalent and derives from the writings of Saint Thomas Aquinas. However, as Bourdieu made clear in two footnotes, he had borrowed the terms from Marcel Mauss's essay 'Techniques du corps', where they had been used to describe the way agents and groups incorporate social imperatives into their deportment and dispositions (see Mauss 1950, pp. 365–86).

'Célibat et condition paysanne' was an early contribution to the growing literature examining the transformation of French rural life, of which the most notable examples were Mallet's *Les Paysans contre le passé* (Peasants against the Past) (1962), Michel Debatisse's *La Révolution silencieuse* (The Silent Revolution) (1963), Henri Mendras's *The Vanishing Peasant* (*La Fin des paysans* 1967), and Morin's *Plodémet* (*Commune en France: la métamorphose de Plodémet* 1967). Much of this literature focused on the efforts of a militant national movement of young farmers to embrace progress and force the pace of modernisation. Mallet's book, for example, examined this movement in the Finistère region. Debatisse was himself the movement's high-profile leader; close to both the PSU and CFDT, his faith in modernisation reflected in part the influence of Teilhard de Chardin's philosophy.[4] Mendras's study placed considerable faith in the ability of these progressive farmers to ensure French agriculture's transition from tradition to modernity, seeing the rural exodus as both an inevitable and beneficial part of this process.

In emphasising the sense of cultural dislocation and anomie which the processes of modernisation were provoking in the Béarnais peasants, Bourdieu's analysis might be read as a necessary foil to such optimistic predictions for the future of the French peasantry. However, it should be noted that Morin's *Plodémet* by no means manifested the kind of uncritical faith in technological progress which Bourdieu's criticisms of its author might have suggested. Indeed, in its emphasis on both the problematic nature of modernisation and the role of women, radio, cinema and pop music as agents of cultural change *Plodémet* had many points in common with 'Célibat et condition paysanne'.

Where Bourdieu's analysis of the rural exodus was perhaps most distinctive was in its debt to the Durkheimian tradition and to the work of Maurice Halbwachs in particular. Bourdieu's emphasis on the clash between the peasant ethos and a set of more modern, urban values owed a considerable debt to Halbwachs's *The Psychology of Social Class* (1958). His focus on shifting patterns of marriage and celibacy recalled an essay of 1935, 'La Nuptialité en France pendant et depuis la guerre' (The Marriage Rate in France during and after the War), in which Halbwachs had shown how the loss of a generation of French males in the First World War had forced women to alter their expectations of who might constitute a suitable marriage partner (in Halbwachs 1972, pp. 231–74). Bourdieu's accounts of the erosion of the collective limits set on ambition and expectation embodied in the peasant ethos, meanwhile, recalled Durkheim's analysis in *Suicide* (*Le Suicide* 1897) of the effect of nineteenth-century industrialisation in breaking down collective limits on desires and ambitions and unleashing an unseemly scramble to satisfy new needs and appetites (1897, pp. 252–3).

In both *Travail et travailleurs …* and *Le Déracinement*, Bourdieu had shown how, in migrating into Algeria's shantytowns and 'resettlement centres', Algerian peasants had been forced to adopt new lifestyles which imposed a set of 'new needs', which had not been a feature of life in their villages. In the 1966 essay, 'Différences et distinctions' (Differences and Distinctions), he traced the fortunes of the French peasants and workers who were migrating from their villages and traditional working-class communities into the *grands ensembles*, the new housing estates springing up around France's major cities and towns. Bourdieu would show that this migration was unleashing an analogous series of new needs and desires, whilst establishing new grounds for class distinctions.

'Différences et distinctions' was just one of three essays Bourdieu contributed to the 1966 volume, *Le Partage des bénéfices: expansion et inégalités en France*. This collective study grew out of a colloquium organised the previous year by Bourdieu and his colleague from the INSEE, the statistician Alain Darbel. The colloquium was a response to the perceived need to examine and correct some of the more extravagant claims made by a whole series of commentators, by means of a detailed empirical and theoretical analysis of the true extent and limitations of the social and economic changes wrought by France's postwar reconstruction. Bourdieu himself contributed three essays to the volume, of which 'Différences et distinctions' was perhaps the most interesting since it sought to formulate general hypotheses about the effects of French economic expansion, focusing on the rural exodus, the mass availability of consumer durables, and the emergence of the socially mobile and highly aspirational *classes moyennes* as a significant social force.

For Bourdieu (1966, p. 119), the increasing expenditure on consumer goods in postwar France was attributable primarily to the rural exodus and 'the victory of the town over the countryside, of urban values over rural values'. Migrants from the countryside had left behind the ascetic ethos of peasant life. They willingly availed themselves of the new consumer credit opportunities to demonstrate their adherence to modern urban values by purchasing a television, fridge or washing machine. However, both the peasants and the workers who had been rehoused in the new housing estates were ill-equipped, mentally and materially, to adapt to this new lifestyle. Isolated and alienated by the erosion of the strong community ties that had characterised their old villages and *quartiers populaires*, they were frequently incapable of meeting 'the new burdens and new needs which are, as it were, implied in modern housing' (p. 121).

Whilst acknowledging that material differences between rich and poor had been mitigated, Bourdieu argued that other kinds of distinction were rapidly taking their place. Brute material differences between rich and poor were increasingly being replaced or supplemented by differences in the way class distinctions were symbolised or signified through different lifestyles. It was no longer a question of simply possessing certain commodities but rather of the quality of those commodities and the manner in which one consumed them; 'a new form of rarity, rarity in the art of consumption and no longer in the goods consumed' (p. 118). The cultural domain was becoming a privileged field for the making of such distinctions. Clothing, language and cultural taste were the most prestigious differences; they symbolised social position most clearly; they could be read as signs of inherent refinement and moral excellence; they were culture transformed into nature. The domain of class distinction, Bourdieu argued, was structured like a language, composed of arbitrary and differential signifying elements; 'symbolic consumption transmutes goods into signifying distinctions' (p. 128).

If the workers and ex-peasants were the losers in this new struggle for social and symbolic distinction, Bourdieu suggested that the upwardly-mobile *classes moyennes*, the sons and daughters of the old petty bourgeoisie, were better placed to take advantage of the new models of social and cultural life. Having more money and less traditional reliance on extended family and social networks, they were well-equipped to benefit from their new environment and the new models of behaviour that accompanied it, creating a lifestyle around the couple or the nuclear family as the primary social unit (p. 122). Further, Bourdieu argued that it was the aspirational *classes moyennes* who were most likely to avail themselves of the increased opportunities for cultural consumption afforded by the mass media. What he termed their 'cultural good will' was a manifestation of their aspirations to social betterment in the cultural domain, a quest for cultural legitimacy 'in the sub-products of learned culture or the substitutes for direct cultural experiences provided by modern means of communication'. Drawing on audience data for television and radio, Bourdieu argued not only that the *classes moyennes* were far more likely to consume 'cultural' programmes on television or radio than the working class, but also that the bourgeoisie were likely to turn up their noses at the 'vulgarisation' of high culture through the 'mass media'. Thus, even the domain of the mass media was, according to Bourdieu, marked by a dynamic of class distinction and division (pp. 126–7).

'Différences et distinctions' was clearly intended as a riposte both to the kind of technocratic vision of a 'society without classes' propagated by Fourastié in *Le Grand Espoir du XXe Siècle* and to

Morin's argument in *L'Esprit du temps* that 'democratisation of high culture is a trend implicit in mass culture' (1962, p. 59), or indeed to Mallet's argument in *La nouvelle classe ouvrière* that improved housing and the spread of mass tourism meant that at the level of lifestyle differences between the working class and bourgeoisie were disappearing (1963, p. 8). It would be wrong, however, to suppose that Bourdieu and his collaborators were the only sociologists to emphasise the uneven nature of French postwar economic growth. In a series of essays from the early 1960s onwards, Lefebvre had examined the sense of alienation suffered by the inhabitants of the new town of Lacq-Mourenx as they confronted the rationalised and reified environment of the *grand ensemble* (1962, pp. 121–30; 1970, pp. 109–28). Like Bourdieu, he highlighted the extent to which the development of 'urban society' was unleashing a set of 'new needs', of manipulated, conditioned, and programmed consumer desires (Lefebvre 1968, pp. 141–2). This increased emphasis on consumption, he maintained, was leading to the transformation of consumer goods into signifiers of wealth and status, 'the consumption of signs, symbols, and signi- fications', hence establishing new grounds for social discrimination (Lefebvre 1967, p. 24). As the possession of consumer durables, such as cars, became more widespread, so the grounds of distinction shifted from the possession or lack of a car to the possession of the 'right' model or marque.

Furthermore, like Bourdieu, Lefebvre had warned that certain sections of the French Left were as guilty of propagating the myths of technocracy as those on the Right (1967, p. 16), whilst underlining 'the dangers of an imprudent use' of the notion of 'the new working class' (1970, p. 125n.3). What did distinguish Lefebvre's analyses from Bourdieu's, arguably making them more convincing, was his typically Marxist emphasis on the economy, locating the motor for increased consumer expenditure in postwar France in what he termed 'neo-capitalism', a post-imperialist stage of economic development characterised by the search for increased surplus profits through constant product differentiation (1967, p. 14n.1). Bourdieu, on the other hand, attributed the advent of mass consumerism primarily to shifts in social morphology, recalling Durkheim's argument in *The Division of Labour in Society* that the fundamental cause of social and cultural change was an increase in 'the number of individuals in relation and their material and moral proximity' (1893, p. 339). However, this surely risked attributing consumerism to a hypostatised 'dialectic of divulgation and distinction' (Bourdieu 1966, p. 118), or some innate tendency in all consumers to seek to distinguish themselves from their peers.

In both 'Célibat et condition paysanne' and his contributions to *Le Partage des bénéfices*, then, Bourdieu had situated himself quite clearly in opposition to the dominant technocratic discourse, providing a timely reminder of the dangers inherent in too hasty assessments of the progress achieved by France's economic recovery. However, his conclusions were not always so very far removed from those offered by the 'massmediologists' whose work he had lambasted in 'Sociologues des mythologies et mythologies des sociologues'. It was in *Photography*, a detailed study of a 'mass' cultural form, that the extent of his distance from the 'massmediologists', particularly at the level of methodology, would become much clearer.

Sociology versus Semiology – Photography

Photography: a middle-brow art is an abridged translation of a collaborative study originally published in 1965 under the title *Un art moyen: essai sur les usages sociaux de la photographie*. The study grew out of Raymond Aron's seminar on 'the image in industrial society' at the CSE and originally included two long essays on the uses of photography in photojournalism and in advertising, by Luc Boltanski and Gérard Lagneau respectively, which were omitted from the English translation. This is a shame since these two essays highlighted the extent to which *Un art moyen*, as a whole, was intended as a direct response to the work of 'massmediologists' such as Barthes and Morin on mass culture, on the profusion of images of modernity and consumerism in cinema, advertising, and the new mass-market magazines such as *Elle*, *Paris-Match*, *L'Express*, and *Marie-Claire*. One of Bourdieu's criticisms of this work had been that it failed to examine empirically either the structures of production or the modes of reception of such images. The two chapters on photography in journalism and advertising in *Un art moyen* were distinctive in basing their findings on *empirical research* amongst the journalists and photographers who actually produced the images in question. Bourdieu himself contributed a general introduction to the volume and two chapters examining photography as a mass leisure practice and the consumption or reception of photographs as a mass cultural form. Again, his findings were rooted in empirical research, namely an ethnographic study of photographic practice in the peasant communities of the Béarn and a more quantitative survey of the practices and tastes of a representative sample of 692 inhabitants of Paris, Lille, and a small, unnamed provincial town. The message of *Un art moyen* seemed clear; empirical sociology was reasserting its pre-eminence over a semiological or structural hermeneutics.

This almost ostentatious 'return' to a tradition of empirical social science was to be made more explicit in Bourdieu's

Introduction, which included a long quotation from that classic text of nineteenth-century experimental science, Claude Bernard's *Introduction à la médecine expérimentale* (1865), followed by a typically Durkheimian assertion that sociology should mimic the methods of the natural sciences (1965, p. 19 [p. 3]). Furthermore, Bourdieu's argument in the opening sections of his contribution to *Photography* followed closely the argument of Durkheim's *Suicide*. In this, Durkheim had attempted to show that even the most apparently personal decision, the decision to take one's own life, was explicable in sociological terms. To this end, he had opened his study by rejecting psychological motivations as an explanation for the incidence of suicide. In *Photography*, Bourdieu was attempting to demonstrate that an area apparently equally governed by personal motivations, namely aesthetic taste, was in fact determined by sociological norms. Like Durkheim before him, then, Bourdieu rejected psychological motivation as an explanation of a social practice, in this case photography (1965, pp. 32–5 [pp. 13–16]). Photography, he argued, was not explicable in terms of 'the anarchy of individual intentions' but needed to be understood by reference to its function within a given social group (p. 39 [p. 19]).

In the peasant communities of the Béarn, for example, Bourdieu argued that photography performed a very specific role within the family unit, namely to record and solemnise collective or family ceremonies and celebrations, weddings, christenings, or reunions. Where Durkheim, in the *Elementary Forms of Religious Life* (1912), had emphasised the role played by rite and festival in cementing group solidarity, Bourdieu argued that photography now supplemented that role: 'If one accepts, with Durkheim, that the function of the festivity is to revitalise and recreate the group, one will understand why the photograph is associated with it, since it supplies the means of solemnizing those climactic moments of social life in which the group solemnly reaffirms its unity' (1965, p. 41 [pp. 20–1]).

In the Béarn, then, photography fulfilled a specific social function; subjects deemed worthy of photographing were limited to certain family and group occasions and the practice of photography was itself conventionally left to a professional photographer. Drawing on Mauss's famous essay *The Gift* (1923–24), Bourdieu argued that the 'wasteful' or 'conspicuous expense' incurred in hiring a photographer served to solemnise the event further. Wedding photographs, for example, became 'an object of regulated exchange' between the peasants, joining the 'circuit of gifts and counter-gifts which the wedding has set in motion' (1965, p. 40 [p. 20]). Any photograph that did fulfil a clear social function or any peasant who set himself up as an amateur photographer, choosing his subjects at will, would meet the collective reprobation of the community for indulging in unjustifiably 'ostentatious behaviour', which 'like a gift which excludes any counter-gift, places the group in an inferior situation' (p. 75 [p. 48]).

Bourdieu used the Béarn, a highly regulated peasant society in which collective limits on acceptable behaviour were at their highest, as a kind of benchmark against which to measure the extent to which photographic practice amongst other social groups was less strictly subordinated to social function, more genuinely autonomous, with any subject considered worthy of being photographed. However, although Bourdieu's research indicated that possession and personal use of photographic equipment was more widespread among the urban classes, he argued that photography's primary role continued to be that of ensuring the integration of the group. Family and group ceremonies remained the principal subjects photographed even amongst the most cultivated members of the bourgeoisie. Indeed, he only identified two groups amongst whom he detected a genuine tendency to transform photography into an autonomous aesthetic activity. The first group was composed of the upwardly-mobile *classes moyennes*, for whom photography, like other mass cultural practices, provided a substitute for more legitimate cultural activities which were beyond their reach (p. 106 [p. 72]). The second group was composed of bachelors; theirs was an '"egoistic" or "anomic"' practice, resulting from a low level of social integration, 'to use the terms with which Durkheim characterises the different types of suicide' (p. 67 [p. 41]).

Bourdieu's analyses in *Photography* were, then, clearly greatly indebted to the Durkheimian tradition and this turn to Durkheim needs to be understood as an attempt to reassert the value of a classically sociological approach against the kind of semiological analyses popularised by Barthes and other 'massmediologists'. However, it would be wrong to suppose that Bourdieu simply applied a pre-formed Durkheimian analytic framework to his analysis of photography. On the contrary, he introduced two major theoretical innovations, in the form of the concepts of 'habitus' and of 'cultural legitimacy', which were to play a central role in his work henceforth. The concept of 'habitus' had, of course, already made an appearance in his work on the Algerian sub-proletariat and the Béarnais peasantry. However, the Introduction of *Photography* provided Bourdieu with an opportunity to formalise the concept for the first time.

Having argued that the practice of photography could not be attributed to personal or psychological motivations, Bourdieu suggested that it was only the concept of a class-determined habitus which could explain the different ways in which social groups practised photography. In an argument which foreshadowed his critiques of structuralism and Sartrean existentialism in his later works of Kabyle anthropology, Bourdieu proposed the habitus as an alternative to explaining social action in terms either of existential free choice, on Sartre's model, or of unconscious submission to structural law, as the structuralists would have it. The habitus described how objective reality, as measured by the statistical chances of a particular course of action meeting with success, became internalised into a structure of dispositions and aspirations, an implicit sense of what could or could not be reasonably achieved, so as to generate a set of objectively determined practices which were experienced, at the subjective level, as free choices. The habitus was determined by class in two ways, firstly because the objective chances of, say, an Algerian sub-proletarian finding a job or the son of working-class parents reaching university were closely correlated with social origin and, secondly, because the collective historical experience of the sub-proletariat or working class had engendered a 'class ethos', a collective sense of 'the field of objective possibilities', of the limits to reasonable or achievable ambition. In the specific case of photography, this 'field of possibilities' became a 'field of the photographable', so that for a Béarnais peasant to photograph anything other than a family or group occasion would be to transgress the limits of the peasant ethos and incur the disapprobation of the whole group (pp. 20–8 [pp. 2–10]). At the core of Bourdieu's argument in *Photography* was his insistence that a mass cultural form such as photography could only be properly understood as long as one grasped the class ethos and habitus through which its practice and appreciation was mediated.

Thus, for example, Bourdieu found that the practice and appreciation of photography amongst the peasantry and working class revealed a concern with the possible function of a photograph above all else, with its role in solemnising weddings, christenings, and so on. Moreover, their judgements on which subjects might make a 'beautiful', 'interesting', 'insignificant', or 'ugly' photograph revealed an analogous functionalist or realist aesthetic at work. Any photograph or potential subject for a photograph which did not fulfil an identifiable function, whether by recording an event of note or capturing a subject, such as a sunset, which was picturesque in the most conventional sense, would meet with incomprehension. According to Bourdieu, this realist or functionalist 'popular aesthetic' reflected an ethos and habitus engendered in conditions of material necessity which ensured that the peasantry and working class were deeply inimical to art for its own sake, to art which served no apparent purpose. The aesthetic judgements of peasantry and working class were in the last instance ethical judgements (p. 123 [p. 86]).

As Bourdieu pointed out, the functionalism of the popular aesthetic represented the antithesis of a genuinely autonomous aesthetic, at least in the terms in which such an aesthetic had been defined in Immanuel Kant's *Critique of Judgement* (1790), the founding text of modern aesthetics. In Kantian terms, the popular aesthetic was an example of 'barbarous taste', of a taste ruled by a direct appeal to the senses or by the application of ethical standards rather than by that 'disinterested' contemplation of 'pure' aesthetic form, which, according to Kant, defined the specificity of the truly aesthetic experience. For Bourdieu, however, such aesthetic 'disinterest', far from guaranteeing the universal validity of aesthetic judgements, as Kant claimed, was itself merely the expression of a typically bourgeois ethos and habitus; it was a disinterest born of a sense of material security and which expressed the certainty of moral and intellectual superiority (pp. 113–34 [pp. 77–94]).

However, if both the popular and the bourgeois aesthetic were equally contingent upon social conditions, it was the bourgeois aesthetic which enjoyed sole legitimacy. To misquote Marx, the 'legitimate' aesthetic was the ruling idea of the ruling class; it was the means by which the bourgeoisie could *naturalise* its political and economic dominance by asserting that such dominance reflected its innate intellectual and moral superiority. Bourdieu's empirical research had shown that the bourgeoisie's practice of photography was as functionalist as that of the peasantry and working class. Nonetheless, when asked to express an opinion on photography's inherent aesthetic value, they typically responded in one of two ways. Since as a class they possessed the greatest knowledge of fine art and high culture, they might use photography as 'an opportunity to actualise the aesthetic attitude, a permanent and general disposition' (p. 95 [p. 65]), judging photography according to a set of autonomous aesthetic criteria which they so notably failed to apply to their own photographic practice. On the other hand, Bourdieu argued, they might simply refuse to accord any aesthetic value to photography at all, suspicious of its popularity and easy accessibility, rejecting any 'fervent attachment to a practice suspected of vulgarity by the very fact of its popularisation' (p.74 [p. 47]). Photography, a mass cultural practice, which fulfilled essentially the same function for all social classes and whose cheapness made it accessible to all, was nonetheless subject to a dynamic of social distinction and snobbery. Whether judging photography according to the criteria of 'legitimate' aesthetics or rejecting it as devoid of aesthetic merit, Bourdieu's bourgeois respondents were all seeking to distinguish themselves from the popular perception of photography, from the stereotypical image of vulgar 'snap-happy' tourists so intent on taking a photograph that they failed to contemplate the beauty of what was on the other side of the lens (pp. 100–1 [pp. 68–9]).

Further, Bourdieu argued that the working class were themselves wholly aware of this dynamic of social distinction, of the extent to which their own cultural practices and preferences were disparaged and considered without legitimacy. Rather as in *Economy and Society* Weber (1968, p. 32) had argued that the thief who went about his business surreptitiously thereby manifested a certain recognition of the de facto legitimacy of the rule of law even as he infringed it, so Bourdieu described how the responses of his working-class research subjects manifested an analogous recognition in transgression (1965, p. 135n.30 [p. 196n.30]). They expressed both an adherence to the norms of the popular aesthetic and an awareness that there existed another more legitimate aesthetic with which their tastes ought to comply. This constituted a 'dual set of norms', forcing, 'the same subject to distinguish between what he likes doing and what he ought to like doing' (p. 121 [p. 84, trans. modified]).

By means of a detailed empirical analysis of one particular mass cultural form and drawing on the classical sociological tradition, Bourdieu had, in *Photography*, sought to refute the claims of the 'massmediologists' regarding the cultural democratisation and homogenisation brought by mass culture. Where *Photography* attempted to demonstrate that class distinctions had a determining force even in the domain of mass culture, *The Love of Art* sought to do the same for the realm of legitimate culture, demonstrating that even in the absence of any formal barriers to access to high culture, in the form of admission fees or cultural provision, the tendency to consume consecrated works of art was closely correlated with social class and education. Once again, this would bring Bourdieu into conflict with contemporary French commentators on culture, particularly those associated with the French New Left.

Inside the Art Gallery

Originally published in French in 1966, *L'Amour de l'art* was reprinted in an extended version in 1969 which broadened the scope of Bourdieu's empirical research to include France and four other European countries, Holland, Spain, Greece, and Poland. It is this extended edition which was translated in 1991 as *The Love of Art*. The book's co-author, the statistician Alain Darbel, used the data from this extensive empirical research into the composition of the European gallery-visiting public to elaborate a statistical model predicting the probability of visiting a gallery according to variables of class, age, sex and education. Bourdieu himself contributed the interpretative commentary on the data, seeking to establish more specifically the way in which such variables affected the tendency to visit an art gallery and hence to consume legitimate culture.

At its simplest level, *The Love of Art* highlighted the very close correlation between social class and education, on the one hand, and the tendency to visit an art gallery, on the other; the cultured bourgeoisie being far more likely to visit art galleries on a regular basis than their working-class counterparts (1969, pp. 35–66 [pp. 14–36]). On a more ethnographic level, Bourdieu noted the frequently rebarbative nature of the gallery environment, the sense of respectful distance it demanded of the public, and the fact that few, if any, concessions were made to those working-class or petty-bourgeois visitors who lacked an intimate knowledge of art and artists. He used these findings to pursue the critique of legitimate aesthetics he had first sketched in *Photography*. Indeed, his very emphasis on the influence of social class and education on the tendency to visit art galleries was in itself a challenge to Kant's notion of the universal nature of disinterested aesthetic contemplation, as well as to what Bourdieu termed 'the ideology of the gift', the notion that the ability to appreciate fine art was an innate 'gift' (p. 90 [p. 54]). The appreciation of a work of art was, he argued, an act of decipherment which demanded possession of the requisite cultural code. Members of the bourgeoisie were far more likely to possess such a code, not simply because of their longer exposure to formal education but also thanks to a more general familiarity with the things of taste and culture, an aesthetic disposition they had acquired in earliest childhood by inhabiting a cultured environment of which high art and culture formed an integral part.

Thus, Bourdieu argued, the aesthetic disposition, measured by the propensity to visit art galleries and hence appreciate fine art, formed part of the bourgeois habitus, constituting a stock of 'cultural capital' which the dominant class could exploit to naturalise and reproduce their dominant status. In attributing aesthetic taste to an innate gift, 'the ideology of the gift' functioned by masking the social determinants of the aesthetic disposition, allowing a socially determined propensity to consume legitimate culture to become a marker of apparently natural intellectual and moral superiority. Further, Bourdieu argued that in a world increasingly subject to the laws of the market, to indulge in the 'disinterested' pleasures of the aesthetic had itself become a luxury, the symbol of the dominant class's objective distance and subjective sense of distinction from the realm of vulgar material necessity inhabited by the dominated classes:

> the privileged classes of bourgeois society replace the difference between two cultures, products of history reproduced by education, with an essential difference between two natures,

one nature naturally cultivated and one nature naturally natural. Thus, the sanctification of culture and art, this 'currency of the absolute' [*cette 'monnaie de l'absolu'*] which is worshipped by a society enslaved to the absolute of currency, fulfils a vital function by contributing to the consecration of the social order. (p. 165 [p. 111, trans. modified])

Bourdieu's critique of the art gallery as an institution, of legitimate aesthetics and of its role in naturalising and reproducing class differences has been discussed at some length by previous commentators on his work. What has attracted less attention, however, is the fact that this critique was carried out not merely at the general level, against legitimate aesthetics per se, but also had a series of more specific targets closely related to the historical context in which Bourdieu was writing.[5] The placement of the phrase 'currency of the absolute' ['*monnaie de l'absolu*'] within inverted commas in the above quotation offers one clue as to whom, more precisely, Bourdieu was seeking to attack here. For *La Monnaie de l'absolu* was the title of the third volume of André Malraux's *Essais de psychologie de l'art* (1948–50).

Malraux had been appointed de Gaulle's culture minister in 1959 and was to remain in post until de Gaulle's fall from power in 1969. He headed a ministry whose mission, according to the decree of July 1959 which had founded it, was to: 'To render humanity's, and above all France's greatest works of art accessible to the greatest possible number of French citizens, to ensure the widest audience for our cultural heritage, and to promote the creation of both the works of art and the spirit which enrich that heritage' (quoted in Loosely 1995, p. 37). This mission had fed into the Fourth Plan (1962–65), which, for the first time, had a commission dedicated exclusively to cultural provision. Malraux himself oversaw the creation of a 'Service des expositions', which mounted a series of high-profile international exhibitions dedicated to such as Picasso or Tutankhamun, and launched an ambitious plan to build 'Maisons de la culture' in France's major provincial towns, completing eight between 1961 and 1968.

Malraux's culture ministry was not the only body involved in an active campaign to achieve a greater measure of cultural democratisation. The postwar period in France had seen the establishment of a series of groups who hoped to bring culture to the masses by staging cultural events in working-class districts or at workplace canteens. Amongst the best-known of these were Joffre Dumazedier and Benigno Cacérès's Peuple et Culture, Jean Vilar's Théâtre national populaire (TNP), and Roger Planchon's theatre in the working-class suburb of Villeurbanne. The campaigning agenda behind these various popular culture movements had been set out in a series of influential studies of the early 1960s, most notably Dumazedier's *Vers une civilisation du loisir?* (Towards a Civilsation of Leisure?) (1962), Jacques Charpentreau and René Kaës's *La Culture populaire en France* (Popular Culture in France) (1962), and Cacérès's *Histoire de l'éducation populaire* (History of Popular Education) (1964). Whilst these movements had initially been associated with the French Left, as Brian Rigby (1991, pp. 132–3) has pointed out, with the installation of Malraux as Minister for Culture in 1959, many of the leading lights of the popular culture movement became co-opted into state-run policies for cultural development. Further, many of these figures were former disciples of Emmanuel Mounier, founder of *Esprit*, the journal which Bourdieu had identified as one of the 'new exchange places' between state planners and the French New Left.[6]

Thus, as in his work on class and mass cultural forms, so in *The Love of Art* Bourdieu set out to demonstrate the objective affinity between a Gaullist minister, such as Malraux, and certain sections of the French New Left. He argued that both Malraux and cultural *animateurs* such as Dumazedier and Cacérès shared the same basic philosophy, the same belief that the ability to appreciate art was somehow natural and that it was therefore sufficient to expose the working class to great works of culture to satisfy their innate 'cultural needs', needs universally shared by every social class. As David Looseley (1995, p. 40) has pointed out, Malraux's adherence to the doctrine of cultural needs implied an almost religious belief in the revelatory power of the work of art; the Maisons de la culture were to be 'cathedrals to culture' in which the masses would 'commune' with high cultural artefacts. This, in turn, implied considerable hostility to the notion that the Ministry of Culture's mission should overlap with that of Education. In *The Love of Art*, Bourdieu noted a similar hostility to providing formal education in the principles of art appreciation in Charpentreau and Kaës's *La Culture populaire en France*, attributing it to their 'class ethnocentrism', to the belief they shared with the *animateurs* of the Maisons de la culture that 'the least cultivated classes (thus the least corrupted by the routinising influence of university education), are predisposed by their state of cultural innocence to receive the most authentic and daring forms of art without prejudice' (1969, p. 151 & p. 152n.21 [p. 103 & p. 172n.21]).

In insisting on the importance of education as a determinant of the propensity to visit art galleries, Bourdieu was thus not simply questioning the supposedly universal criteria of legitimate aesthetic judgement, he was also contesting the notion that universal 'cultural needs' could be satisfied by overcoming 'the mere physical inaccessibility of the works of art' (p. 151 [p. 103]). These apparently universal cultural needs were in fact the product of a particular education: '"cultural need" ... , in contrast to "primary needs", is the result of education' (p. 69 [p. 37]). Simply removing the formal obstacles to legitimate culture, whether economic or geographical, was, Bourdieu argued, to overlook the whole series of cultural, educational and class determinants affecting the propensity to consume legitimate culture:

> The exhibitions of paintings in Renault factories or the plays put on for the workers of Villeurbanne are experiments which can prove nothing since they make the very object of the experiment disappear, by taking as already solved the problem that they claim to solve, namely that of the conditions of cultural practice as a deliberate and regular enterprise; but at all events they manage to convince those who undertake them of the legitimacy of their enterprise. (p. 151 [p. 103, trans. modified])

Invoking 'the sobering realism of statistics', Bourdieu challenged the assertions of those commentators who saw in 'the success of the Picasso or Tutankhamun exhibitions' the 'indications of a democratisation of access to culture' (pp. 131–2 [pp. 87–8]). Similarly, he marshalled a set of statistics which suggested that the Maisons de la culture had conspicuously failed to attract a significantly broader cross-section of society than traditional museums or art galleries.

Against the claims of Malraux and the *animateurs* of the popular culture movement, then, Bourdieu concluded that it was only through the extension of formal education, and not through 'direct cultural action', that a lasting disposition towards aesthetic appreciation could be engendered amongst all social classes. A 'rational pedagogy' was required, in which the principles of art appreciation would be rendered entirely explicit rather than being left to the diffuse and implicit operations of a 'traditional education', which 'abandoned the task of cultural transmission to the family', hence to a class-determined habitus (pp. 104–5 [pp. 65–6]). At present, Bourdieu argued, by failing to provide 'rational' formal instruction in art appreciation, the school was abdicating its responsibility to transmit equal amounts of knowledge to all, regardless of their social background:

> the educational institution abdicates its very specific responsibility, namely the power of exercising the continued and prolonged action, methodical and uniform, in other words universal or tending towards universality, which (to the great scandal of the holders of the monopoly on cultivated distinction) is alone capable of *mass-producing* competent individuals equipped with the schemes of perception, thought and expression which are the condition for the appropriation of cultural goods, and with the

generalised and permanent disposition to appropriate them. (p. 106 [p. 67, trans. modified])

In arguing that it was the school which both had the duty and the unique capacity to ensure 'universal' access to knowledge, art, and culture, Bourdieu revealed a striking residual faith in the ideals of the French republican tradition. In an essay of 1903, 'Secular Morality', Durkheim had argued that, following the secularisation of French education under the Third Republic, morality needed to be taught not in accordance with the religious principles which had prevailed under the previous 'traditional pedagogic system', but rather in accordance with a set of entirely rational principles, 'an entirely rational moral education'. Only then could French education fulfil its properly republican function of ensuring that France's citizens enjoyed universal access to knowledge and culture, regardless of their religious persuasion or social background (Durkheim 1903). In *The Love of Art*, Bourdieu was rehearsing Durkheim's classically republican argument with reference now to high or legitimate culture rather than to religion. Just as at the turn of the century Durkheim had argued for an education based on the narrow, partial interests of one revealed religion to be replaced with an educational system based on purely 'rational', hence 'universal' principles, so sixty years later Bourdieu demanded that the teaching of art appreciation be 'secularised', organised along 'rational' principles rather than being left to the vagaries of social background or class determined habitus. It is in the context of this drive to secularise and rationalise the teaching of legitimate culture that it is possible to understand Bourdieu's decision to use of a series of religious metaphors to describe the forms of 'belief', 'communion', 'consecration', 'enchantment', 'sacralisation', or 'revelation' which characterised the aesthetic experience as traditionally understood.

However, if the republicanism implicit in Bourdieu's proposals for change appeared consistent with his theoretical debts to Durkheim, it seemed strangely at odds with the radicalism of his critique of legitimate culture. His analysis of the way the aesthetic ideology of the bourgeoisie masked the historical conditions of its inculcation, for example, seemed to owe more to Marx than to Durkheim, representing an extension of the Marxist theory of commodity fetishism to the domain of cultural consumption. Moreover, at times in *The Love of Art* Bourdieu had suggested that the content of legitimate culture was itself entirely culturally arbitrary, the product of a historically determined process of familial and formal education whose sole objective function was to naturalise and reproduce class divisions: 'Inasmuch as it produces a culture (*habitus*) which is simply the interiorisation of the cultural arbitrary, family or school upbringing has the effect, through the inculcation of the arbitrary, of masking ever more completely the arbitrary nature of the inculcation' (1969, p. 162 [p. 109]). If legitimate culture were merely the expression of the ethos of the dominant class and hence culturally arbitrary, there seemed no obvious reason to urge the dominated classes to learn to appreciate such culture. On the contrary, a more consistent course of action would seem to be to call for a radical assault on the bourgeois values transmitted through legitimate culture whilst fighting for its replacement by a more 'authentic' set of cultural forms.

Bourdieu's approach throughout his early work on class and culture had seemed remarkably consistent. He had sought to make sense of the changes sweeping through French society in the 1960s by staging a return to the classical French sociological tradition, personified by Durkheim and his collaborators such as Mauss and Halbwachs, hence challenging the analyses of such changes offered by technocrats and 'massmediologists' alike. However, there seemed to be an inconsistency at the heart of *The Love of Art*, a contradiction between the radicalism of Bourdieu's critique of the 'arbitrary' nature of legitimate culture and the reformism, even republicanism, of his proposals for change. This apparent contradiction centered on the role he attributed to education.

Between 1964 and 1970, Bourdieu published four books and numerous articles on French higher education. In a natural extension of his contemporaneous studies of mass culture, these studies challenged those on Left and Right alike who took the massive expansion in the sector in postwar France as evidence of the straightforward 'massification' or 'democratisation' of the universities. These studies have often been interpreted as implying a straightforward rejection of the republican vision of education, emphasising instead the role of the universities in reproducing and legitimising class divisions. However, as the next chapter will demonstrate, a closer reading of his work on class and education reveals that, as in his critique of legitimate culture in *The Love of Art*, Bourdieu's criticisms of French education coexisted, in a perhaps contradictory way, with a residual belief that, suitably reformed, French universities might indeed realise the duties with which the republican tradition had charged them, namely to ensure democratic and universal access to knowledge and culture.

Notes to Chapter 2

1. On the links between technocracy, the CFDT, and the PSU, see Hamon and Rotman (1984).
2. On Mounier, *Esprit*, and Teilhard de Chardin, see Hellman (1981, pp. 218–20). For a more detailed Bourdieusian analysis of postwar French

sociology and the New Left, see Boltanski (1982, pp. 155–304) and
Bourdieu (1976).

3. In the Introduction to *Algérie 60*, Bourdieu (1977, p. 8) stated that it
was only lack of time that had prevented him from publishing his work
on the Béarn alongside his analyses of the erosion of traditional Algerian
society to provide a single study of analogous processes of modernisation
and rationalisation.

4. On the young farmers' movement, see Wright (1964, pp. 143–82).

5. One notable exception is Brian Rigby, to whose analyses I remain greatly
indebted (see Rigby 1991).

6. On the popular culture movement's links to Mounier, see Rigby
(1989).

CHAPTER 3

Universities

In the last chapter, we saw that Bourdieu appeared to be embracing a peculiarly French republican vision of the role of formal education in ensuring universal access to knowledge and high culture. Echoing Durkheim's support at the turn of the century for a secular and 'rational' education, Bourdieu's calls, in *The Love of Art*, for the establishment of a 'rational pedagogy' to replace a class-based 'traditional' form of education suggested a strong residual faith in the classical ideals of the French republican tradition. However, if this republicanism seemed strangely at odds with the radicalism of Bourdieu's critique of legitimate culture, it seemed directly to contradict his assault on the myths of republican education in studies such as *The Inheritors* (*Les Héritiers* 1964) or *Reproduction* (*La Reproduction* 1970).

The studies of French higher education which Bourdieu published between 1964 and 1970 appear primarily concerned not with education as a positive force for social change, but rather with its role in reproducing and legitimising existing class divisions. In his native France, Bourdieu's work on education tends to be interpreted as part of an essentially Marxist or *marxisant* assault on 'bourgeois ideology', epitomising the spirit of contestation which marked the years immediately preceding and succeeding the events of May 1968 (Collectif 'Révoltes Logiques' 1984; Ferry and Renaut 1987). Indeed, the apparent force of Bourdieu's critique of the class bias of French higher education has led many critics to suggest a close affinity between his work and the analyses of class, education and culture offered by Marxists such as Antonio Gramsci or Louis Althusser (Hall 1977; Garnham 1980; Eagleton 1991; Moi 1991; Alexander 1995). This suggestion of an affinity with the Marxist tradition has frequently been accompanied by the accusation that Bourdieu places an excessive emphasis on education's role in social reproduction, offering, in Richard Jenkins's words, a 'static or unchanging picture' of the educational field (1992, p. 121).

This chapter will argue that such assessments of Bourdieu's writings on education are mistaken. Whatever their similarities, simply to conflate Bourdieu's and Althusser's theories of the role of education is to risk overlooking the very real differences which separate the two.[1] In particular, it is to overlook the key importance of the Durkheimian tradition to Bourdieu's work on education and the extent to which that work remained marked by an always ambivalent, sometimes contradictory relationship to the French republican tradition. Further, excessive emphasis on Bourdieu's interest in education's role in social reproduction can obscure the fact that all his major works on French higher education were attempts to describe the sector at a time of rapid change, occasioned by a massive increase in student numbers. In the academic year 1949–50, the number of students in French higher education was 136,744. By 1959–60 this had almost doubled to 202,062, and by the end of the 1960s it had more than quadrupled to 615,300. This expansion represented a significant increase in the percentage of the French population entering higher education, bringing new categories of student, notably large numbers of women, into French higher education for the first time. Such changes might be seen as symptomatic of a shift in the nature and function of French higher education, a shift consistent with France's need, under late capitalism, for an increasing number of technically and academically trained workers to manage and plan continued economic growth.

Ernest Mandel argues that under late capitalism the role of higher education is redefined, being 'no longer to produce "educated" men of judgement and property – an ideal which corresponded to the needs of freely competitive capitalism – but to produce intellectually skilled wage-earners for the production and circulation of commodities' (1975, p. 261). As Bourdieu put it in *The Inheritors*, French universities were experiencing problems in adapting to this new role since they continued to function according to a 'traditional' rather than a 'rational' logic: 'the logic of a system which, like the French system in its present form, seems to serve traditional rather than rational ends and to work objectively to fashion men of culture rather than men with a trade' (1964b, p. 88 [p. 58]). For Mandel (1975, pp. 260–1), the difficulties faced by universities reflected 'the crisis of the classical humanist university', occasioned not only by the 'excessive number of students', the 'backwardness of material infrastructure', and 'changes in the social background of students', but also by '*directly economic* reasons specific to the nature of intellectual labour in late capitalism; the constraint to adapt the structure of the university, the selection of students and the choice of syllabuses to accelerated technological innovation under capitalist conditions'.

This chapter will argue that Bourdieu's work on higher education, the books and articles he published on the subject between 1964 and 1970, as well as his retrospective study of the French academic field of the 1960s, *Homo Academicus* (1984), are best understood as highly influential interventions into contemporary debates sparked by this 'crisis of the humanist university' or attempts to analyse that 'crisis' post hoc. As with his analyses of postwar French culture and society, Bourdieu's studies of education would challenge what he saw as the hasty conclusions drawn by contemporary commentators both on the Right and the French New Left. Bourdieu would emphasise the need to look beyond the overall expansion in student numbers to grasp the way in which this apparent democratisation of higher education was mediated through the habitus of students, hence reproducing older distinctions of class and gender in subtle new forms. This chapter will examine Bourdieu's analysis of this process in his three book-length studies, *The Inheritors*, *Academic Discourse* (*Rapport pédagogique et communication* 1965), and *Reproduction*, before moving to an analysis of his retrospective account of the shifts in the French academic field in the years leading up to the events of May 1968 in *Homo Academicus*.

The Limits of Democratisation

At their simplest, the books and articles Bourdieu published on education between 1964 and 1970 can be read as attempts to challenge those who pointed to the increase in the number of students entering French higher education as evidence of the advent of an era of educational democracy. As such, they complemented his concurrent work on mass culture, reflecting his scepticism regarding the dawning of an age of cultural homogenisation. Thus, in *The Inheritors*, Bourdieu produced official data for the student intake of 1961–62 which indicated that, despite the expansion in overall access to higher education, older distinctions of both class and gender continued to play a determining role in academic success.

In terms of class, Bourdieu demonstrated that whilst the son of a farm worker had less than one chance in a hundred of reaching university, the son of a member of the liberal professions had more than eighty chances in a hundred (1964b, p. 11 [p. 2]). Following a line of argument familiar from his work on both the Algerian sub-proletariat and photography, he argued that the low objective chances of lower-class children entering higher education were internalised into their habitus, into an implicit sense of what did or did not constitute an objectively possible future, at once a subjective disposition and a class 'ethos' which encouraged such children to rule out university as a 'practical possibility'. If everything in the dispositions, social trajectories and expectations of immediate family and friends militated against children from lower-class backgrounds considering university as a 'practical possibility', the reverse was true for children from a privileged background:

Even if they are not consciously assessed by those concerned, such substantial variations in objective educational opportunity are expressed in countless ways in everyday perceptions and, depending on the social milieu, give rise to an image of higher education as an 'impossible', 'possible' or 'natural' future, which, in turn, plays a part in determining educational vocations. (1964b, p. 12 [pp. 2–3])

This process of internalising a socially determined sense of what constituted an 'impossible', 'possible', or 'natural' future was also manifest in the persistence of gender inequalities within the university. As Bourdieu demonstrated, since the mid-1920s and a fortiori since the Liberation, the proportion of female students had increased dramatically, from 3.2 per cent of the total in 1900–01 to 21.0 per cent in 1925–26 and 41.6 per cent in 1961–62, provoking 'a major cultural transformation' in French universities (1964b, p. 129 [p. 107]). However, he argued that the immense majority of these female students, 63.3 per cent, had ended up in the Arts faculties rather than the more prestigious law or medical faculties and were 'relegated' within the Arts faculty to the least prestigious disciplines, foreign languages, sociology or psychology rather than philosophy or French literature (pp. 17–18 [p. 6]). This process of relegation, Bourdieu maintained, reflected the way a collective sense of what represented a reasonable ambition for a young woman, 'the influence of the traditional model of the division of labour (and the distribution of "gifts") between the sexes' had been internalised into the habitus of female students and their parents to influence their 'choice' of course (p. 17 [p. 6]). He suggested that this gendered habitus was articulated with class inasmuch as female students from more privileged backgrounds were statistically more likely to avoid 'relegation', demonstrating greater confidence in their own abilities by choosing more prestigious courses.

By 1970 and the publication of *Reproduction*, Bourdieu had access to a wider range of data on the gender and social origins of French students and was able to achieve a broader perspective on the nature of the changes he had examined in *The Inheritors*. Whilst acknowledging that between 1962 and 1966 the chances of entering higher education had risen for all social classes, he argued that it was the privileged classes who had benefited most from this development. There had been 'an *upward translation* of the structure of the educational chances of the different social classes', so that where in 1961–62 the possibility of the son of an industrialist reaching university had been 'a probable future', by 1965–66 it had become 'a banal future' (1970, pp. 261–2 [pp. 224–6]). The rise in the chances of the son of a manual worker from 1.5 to 3.9 in a hundred over the same period, however, had not been enough to 'modify the image of higher education as an unlikely, if not "unreasonable" future' (p. 263 [p. 227]).

Furthermore, if certain fractions of the *classes moyennes* could now consider higher education to be 'a normal possibility', this suggested that attending university was rapidly losing its distinctive or rarity value. New grounds of distinction were being produced within the university itself and Bourdieu produced a set of statistics to show that as new categories of student gained access to higher education so the disciplines they chose, primarily in the faculties of science and arts, began to lose prestige. The goalposts were shifting and bourgeois students increasingly abandoned the faculties of science and arts in favour of more prestigious disciplines, medicine, law, or the highly selective *grandes écoles*. As Bourdieu put it, absolute exclusion from higher education was being replaced as the primary means of social and educational distinction by relegation into less prestigious disciplines (pp. 264–7 [pp. 228–31]). A similar process was affecting female students, whose increased chances of access concealed their continuing relegation 'to certain types of study (arts subjects in the main), the more so the lower their social origin'. Although more women were now entering the labour market, Bourdieu argued that the entry of women into a profession such as teaching frequently coincided with that profession's relative loss of prestige (p. 215 [pp. 182–3]).[2]

A central concern of *The Inheritors* and *Reproduction* had, therefore, been to highlight the limits of democratisation and the extent to which older distinctions of class and gender continued to play a decisive role in education, even as they were being reproduced in new forms. However, Bourdieu's emphasis on the way questions of class and gender continued to influence students' experiences of higher education was not merely meant as a riposte to those, such as Jean Fourastié (1963, pp. 283–4), who pointed to the overall increase in student numbers as evidence of the democratisation of higher education. On the contrary, Bourdieu devoted an entire chapter of *The Inheritors* to refuting the analyses of the universities being offered by a range of commentators associated with the French New Left, most notably Marc Kravetz, general secretary of the Union nationale des étudiants de France (UNEF), a students' union close to the PSU and CFDT.

Student Politics

As Luc Boltanski has shown, between 1963 and 1965 UNEF commissioned a series of sociological studies into what they termed 'the student condition [*la condition étudiante*]' (1980, pp. 358–64). UNEF argued that the transformation of French universities from 'liberal' to 'technocratic' institutions under 'neo-capitalism' meant that students were enduring an alienating and exploitative 'condition' which placed them in objective alliance with Serge Mallet's 'new working class' (Kravetz 1964; Griset and Kravetz 1965). In an article of 1967, André Gorz drew on UNEF's analyses to provide a classically Gramscian diagnosis of the situation. There was a contradiction, he argued, between the knowledge and autonomy a neo-capitalist economy demanded of its workforce and the archaic, hierarchical structures of French society. This contradiction, expressed in the frustrations, alienation, and exploitation suffered by students and the 'new working class' alike was melding these two groups into a new 'hegemonic' force, in the Gramscian sense, *'un nuovo blocco storico'* [in Italian in original] endowed with revolutionary potential (Gorz 1967). Gorz's analysis was later to be echoed by Alain Touraine in his explanation of the events of May 1968, *Le Mouvement de mai ou le communisme utopique* (1968).

Drawing on quantitative and ethnographic surveys of the student experience conducted by colleagues at the CSE, Bourdieu challenged the notion, so central to the analyses of the New Left, that 'the student condition is a single, unified, or unifying one' (1964b, p. 24 [p. 13, trans. modified]). He demonstrated the low level of integration amongst the student body and the extent to which students remained divided by class, particularly in their cultural tastes and practices. He argued that the male offspring of the Parisian bourgeoisie had a very different experience of university to that of female or working-class students. The complaints of female and working-class students about the inadequacies of French universities reflected genuine worries about their modest future job prospects, about the ability of 'a teaching little changed in its methods or sometimes even its content, ill-suited to the expectations and interests their background has given them', to prepare them for 'an occupational future with which they have a more realistic concern' (1964b, p. 78 [p. 53]). The apparently radical political affiliations of Parisian male bourgeois students, on the other hand, merely manifested the self-assurance and dilettantism which was intrinsic to their habitus. Membership of a radical left-wing 'groupuscule' was merely the means for the confident bourgeois student 'symbolically to consummate their break with their family background in the least costly and most scandalous way' (p. 70, [p. 47]). Hence, for Bourdieu, talk of a unified 'student condition' or of the students as a potential political force was mistaken. The student body remained divided along class and gender lines whilst any apparent politicisation was essentially a superficial, transitory phenomenon.

If Bourdieu's position contrasted strongly with the classically Gramscian analyses of the French New Left, with their emphasis on the creation of 'new historic blocks' and 'new hegemonic forces', it was also notably distinct from Althusser's position at this moment. In 1964, Althusser published an article entitled 'Problèmes étudiants' (Student Problems) in the French Communist Party journal, *La Nouvelle Critique*. A response to the successful 'week of action' organised by UNEF in November 1963, Althusser's article was even more scathing of student politics than Bourdieu. Althusser instructed the students to respect their lecturers and busy themselves with the acquisition of 'scientific' knowledge which could then be placed at the service of the genuine revolution. He reminded them that the liberal values of the university were both inherently valid and potentially scientific, 'not, as some now dangerously claim, reducible to *bourgeois individualism*, but authentic scientific values' (Althusser 1964, p. 86).

It would not be until the publication of his 1970 essay 'Ideology and Ideological State Apparatuses' that Althusser would identify education's particular role in social reproduction or acknowledge the importance of struggles in the educational field in an attempt, according to Jacques Rancière (1974), to excuse both his earlier hostility to student politics in 'Problèmes étudiants' and the French Communist Party's inaction in May 1968. Significantly, in their 1971 *L'École capitaliste en France*, the Althusserians Christian Baudelot and Roger Establet credited Bourdieu with having been the first to identify education as 'the ideological apparatus number one', even as they criticised his work for relying on essentially 'ideological' and 'unscientific', hence insufficiently Marxist, premises (1971, pp. 313–14).[3] It was Bourdieu's greater sensitivity to the problems faced by French students which surely explains how, despite his criticisms of the UNEF position, *The Inheritors* was able to have such an influence on the nature of the students' demands in the run up to May 1968.[4]

From the 'Organic' to the 'Critical'

With hindsight, it is possible to see that, in *The Inheritors*, Bourdieu considerably underestimated the importance of the student movement. However, this should not obscure the prescience of his analyses of the difficulties French universities faced in struggling to adapt to the new demands being placed on them. This was particularly true of the collaborative study Bourdieu published in 1965, *Academic Discourse*, which diagnosed a failure of communication between lecturers and their students, focusing on the physical and symbolic distance that separated the two parties. Given that the events of May 1968 were characterised, as one commentator puts it, by 'confrontations and attempts at communication across generational and class lines that were unparalleled in [France's] recent national life', such analyses were prescient indeed (Turkle 1978, p. 3).

Two years after the events of 1968, in *Reproduction*, Bourdieu returned to this communicative breakdown, linking it more explicitly to the changes in the morphology of the student body which had accompanied a period of rapid expansion. It was these morphological changes, he argued, which had provoked 'the present crisis of the educational system, i.e. the dislocations and breakdowns which affect it as a communication system' (1970, p. 115 [p. 91]). Under a previous, more selective system of university entry, only those equipped with a high level of linguistic and intellectual ability had been able to gain a place. As the universities expanded to accept virtually any student of bourgeois origin, bourgeois students became 'under-selected', no longer possessing the abilities required at a previous stage of the system's history. Similarly, under a more selective system, only the very best students from the working class or *classes moyennes* could hope to reach university; they were 'over-selected'. Increasingly, however, the 'the corrective effect of over-selection', which had compensated working- and lower middle-class students for their lack of the requisite cultural and linguistic habitus, was waning and the potential for linguistic and cultural misunderstandings increased (p. 114 [p. 91]).

Bourdieu maintained that it was this communicative crisis, provoked by the changing composition of the student body, that had revealed French universities' role in legitimising class distinctions. Previously, this role had passed unnoticed since the linguistic and cultural aptitudes demanded by the university were in accord with a body of almost exclusively bourgeois male students. With the arrival of categories of student previously excluded from the system that 'pre-established harmony' had broken down to reveal the social and cultural presuppositions upon which it had rested:

> The situation of nascent crisis provides the opportunity to discern the hidden presuppositions of a traditional system and the mechanisms capable of perpetuating it when the pre-requisites of its functioning are no longer completely fulfilled. It is at the moment when the perfect attunement between the educational system and its chosen public begins to break down that the 'pre-established harmony' which upheld the system so perfectly as to exclude all enquiry into its basis is revealed. (1970, pp. 124–5 [p. 99, trans. modified])

This passage pointed to two distinct, if related lines of enquiry, both of which were linked to the crisis in French higher education. The first was to use that crisis to uncover the 'hidden presuppositions' of a 'traditional' system of education. The second was to explain how that system was able to perpetuate itself when 'the prerequisites of its functioning' had been undermined.

Book One of *Reproduction*, with its series of numbered propositions, sought to formalise the hidden presuppositions of a traditional system of education into what Bourdieu termed 'a theory of symbolic violence'. 'Symbolic violence' was defined as 'the imposition of a cultural arbitrary by an arbitrary power' (1970, p. 19 [p. 5]). The cultural arbitrary in question was legitimate culture, considered arbitrary since it reflected not a universal disposition but rather one specifically attuned to the bourgeois habitus, that repository of a set of values, aptitudes, modes of thought and speech picked up from earliest childhood through being socialised in a cultured, literate, bourgeois environment. It was these values, Bourdieu maintained, which were rewarded in the traditional French educational system, which placed such emphasis on oral skills and rhetorical flair, valuing form over content and encouraging a kind of intellectual 'dilettantism' consistent with the linguistic and cultural capital which constituted the bourgeois habitus. Passing off such arbitrary but socially determined skills and values as universal, objective measures of intellectual ability, the education system both legitimised and reproduced existing class divisions.

It was perhaps in acknowledgement of the dangers of determinism or functionalism implicit in this theory of symbolic violence that Bourdieu emphasised that the theory should be understood as enjoying a reciprocal relationship to the specific historical circumstances which had engendered it, the theory being applied to those circumstances and the circumstances influencing the nature of the theory according to a process of 'mutual rectification' (p. 9 [p. ix]). As he put it:

in order to escape the illusion inherent in a strictly functionalist analysis, we must re-insert the state of the system grasped by our survey into the history of its transformations. The analysis of the differential reception of the pedagogic message presented here makes it possible to explain the effects which transformations in the receiving audience exert on pedagogic communication and to define by extrapolation the social characteristics of the audiences corresponding to the two limiting states of the traditional system – what might be called the *organic* state, in which the system deals with an audience perfectly matching its implicit demands, and what might be called the *critical* state, in which, with the changing social make-up of the system's clientele, misunderstanding would eventually become intolerable – the state actually observed corresponding to an intermediate phase. (pp. 113–14 [p. 90, trans. modified])

If the changing nature of the student body had signalled the end of the previous 'organic' stage in French higher education without leading to its complete collapse, this was, Bourdieu argued, largely because of the system's 'relative autonomy'.

Both Luc Ferry and Alain Renaut (1985) and Jeffrey Alexander (1995) see this emphasis on the role of a 'relatively autonomous' education system in the 'reproduction' of the status quo as evidence of the direct influence of Althusser on Bourdieu's work. However, the notions of social reproduction and relative autonomy are neither exclusively Althusserian nor even exclusively Marxist in origin. Indeed, Bourdieu himself indicated that he was drawing on Durkheim's rather than Althusser's understanding of these two concepts. In his lectures in the history of education in France, posthumously published as *The Evolution of Educational Thought* (*L'Évolution pédagogique en France* 1938), Durkheim had linked French education's relative autonomy to its ability to reproduce its own corps of teachers and professors. Bourdieu radicalised this insight, arguing that the system's relative autonomy was related not merely to its role in the reproduction of university staff, but also to its role in the reproduction of the class structure as a whole; relative autonomy was 'the necessary and specific condition for the performance of its class functions' (1970, p. 238n.22 [p. 215n.21]). As he explained:

> Durkheim, in conceiving the relative autonomy of the educational system as the power to reinterpret external demands and take advantage of historical opportunities so as to accomplish its internal logic, at least obtained the means of understanding the tendency to self-reproduction which characterises academic institutions ... . The fact remains that if one fails to relate the relative autonomy of the educational system to the social conditions of the performance of its essential function, one is condemned ... to put forward a circular explanation of the relative autonomy of the system in terms of the relative autonomy of its history and vice versa. (pp. 231–3 [pp. 195–7])

Echoing Durkheim's *The Evolution of Educational Thought*, Bourdieu argued that the peculiarities of the French system, its emphasis on form over content, its reward for rhetorical skill and dilettantism, even its excessive interest in training the next generation of teachers and lecturers, could be traced back to the influence of the Jesuits from the Renaissance onwards (1970, pp. 171–84 [pp. 142–52]). If these anachronistic forms of learning could survive, despite their lack of suitability given the composition of the contemporary student body and the demands of a modernising French economy, it was because of the system's relative autonomy, 'the reward [*la contrepartie*]' for its ability to reproduce the dominance of the dominant classes (p. 186 [p. 153]).

As the Althusserians Baudelot and Establet pointed out, Bourdieu's conception of reproduction and relative autonomy had little in common with the Marxist understanding of such terms. According to Marxist theory, under capitalism it is not divisions of class or caste which are reproduced, as Weberian or Durkheimian social theory might suggest, but antagonistic relations of production (Baudelot and Establet 1971, pp. 284–5). For Baudelot and Establet, Bourdieu's analyses were functionalist; they failed to recognise the educational system as a site of contradictions and antagonisms which engendered class struggle, enabling the system's relative autonomy to have a 'retroactive' effect on the wider social and economic fields rather than simply contributing to society's unproblematic reproduction (1971, p. 316).

It was not only the concepts of reproduction and relative autonomy which Bourdieu borrowed and adapted from the Durkheimian tradition. His analysis of the breakdown in the 'pre-established harmony' between lecturers and the student body, occasioned by the expansion in student numbers, recalled the essays and lectures on education Durkheim had given at the turn of the century, a time of equally rapid change in the French university sector. Durkheim argued that the emphasis on dilettantism and rhetorical flair, which the traditional French educational system had inherited from the Jesuits, was out of step with the needs of the secular, democratic Third Republic, in which students from a broader range of social backgrounds were gaining access to education. As he put it, 'our traditional system of education is no longer in harmony with our ideas and our needs' (Durkheim 1956, p. 103). French education was undergoing one of those 'critical periods when it is urgently necessary to put a scholastic system back in harmony with the needs of the time' (pp. 104–5). By establishing a 'rational pedagogy', substituting 'rational methods ... for the old routines', the French education system could realise its democratic, republican mission (p. 113).

Throughout *The Inheritors*, *Academic Discourse* and *Reproduction*, Bourdieu had understood the 'crisis' affecting French higher education in terms of the contradiction between a system which, 'while remaining traditional ... operates in a cultural context dominated by rationalisation and the values of rationality' (1965a, p. 13n.1 [p. 29n.1]). In both *The Inheritors* and *Academic Discourse*, he had echoed Durkheim in calling for the establishment of a 'rational pedagogy' as a means to facilitate genuine democratisation: 'Any real democratisation presupposes ... that the area of what can be rationally and technically acquired by methodical learning be enlarged at the expense of what is irreducibly abandoned to the random distribution of individual talents, that is, in fact, to the logic of social privileges' (1964b, p. 111 [p. 73]). School would be 'the royal road to the democratisation of culture', if only it were reformed along rational lines (1964b, p. 35 [p. 21]).

Such apparently straightforward statements of faith in the classical ideals of French republican education were, however, qualified in the very last sentence of *The Inheritors*. Here Bourdieu stated that if 'a truly rational pedagogy ... would help to reduce inequalities in education and culture', this 'would not be able to become a reality unless all the conditions for a true democratisation of the recruitment of teachers and pupils were fulfilled, the first of which would be the setting up of a rational pedagogy' (1964b, p. 115 [p. 76]). At the opening of *Reproduction*, Bourdieu seemed to reject the notion of a rational pedagogy altogether, dismissing the 'utopian character' of the 'hypothesis of the democratisation of education through the rationalisation of pedagogy' (1970, p. 69 [p. 53]). Later in the text, he would advocate the establishment of an 'explicit pedagogy' but argue that 'only a school system serving another system of external functions and, correlatively, another state of the balance of power between the classes, could make such pedagogic action possible' (pp. 162–3 [p. 127]).

Bourdieu's relationship to the Durkheimian tradition and the republicanism which it embodied was thus rather ambivalent. Clearly his analyses owed much to Durkheim's earlier work on French education yet he remained sceptical regarding the possibility of reforms to the system ever being able to make it achieve its stated democratic objectives. A similar ambivalence characterised his critique of the legitimate culture dispensed in the universities. At times, he suggested such culture had no intrinsic value over and above its role in legitimising and naturalising class distinctions. At other times, he implied that it was merely 'the relation to culture' rewarded by the education system which determined this role by demanding that culture should be consumed with a refinement and ease which was the preserve of those who had been acculturated in a bourgeois environment (1970, p. 116 [p. 130]). As Raymond Aron pointed out, discussing *The Inheritors*, Bourdieu 'always seemed to leave his readers with a choice between two interpretations of his critique of the universities; did he wish that everyone should be able to accede to learned culture or did he judge that culture itself severely, at least in the form given to it by traditional universities?' (1968, p. 80).[5]

If it seemed unclear whether Bourdieu's critique was directed at legitimate culture per se or merely the manner in which such culture was taught, it also seemed strange that he should be paying so much attention to this kind of literary culture at a time when, by his own admission, arts faculties were increasingly losing their prestige to other disciplines. On the one hand, Bourdieu had argued that as new categories of student entered the university they were relegated to the arts and humanities, subjects which were losing their prestige as bourgeois students gravitated to the medical and law faculties or the *grandes écoles*, where selection still applied. On the other hand, in attributing the high rate of educational success amongst the bourgeoisie to a cultured, literate habitus closely attuned to the classical humanist values rewarded by the education system, he seemed to imply that the prestige traditionally attached to the humanities had remained largely unchanged. Further, Bourdieu's research was conducted exclusively amongst students in the arts and humanities. As he put it in *The Inheritors*, this was because, 'the influence of social origin is most clearly manifested in the teaching of the humanities' (p. 19 [p. 8]). By his own admission, then, the influence of social origin was most evident in those disciplines which, under the new academic hierarchy, had little prestige, whilst his analysis of the contribution of the bourgeois habitus to the academic success and reproduction of the bourgeoisie was based on empirical research conducted amongst students in disciplines that were slipping to the bottom of the academic hierarchy.

In *Reproduction*, Bourdieu insisted that the newer scientific disciplines could also play a role in social distinction, arguing that: 'it would be naive to suppose that the function of social distinction performed by the cultivated relation to culture is exclusively and eternally attached to "general culture" in its "humanistic" form. The glamour of econometrics, computer science, operational research or the latest thing in structuralism can serve ... as an elegant ornament or an instrument of social success' (1970, pp. 156–7 [p. 123]). However, if 'legitimate culture' was now to embrace everything from computing to Corneille, the concept surely risked losing all its specificity and explanatory force. It would only be with the publication of *Homo Academicus* in 1984 that Bourdieu would address in more detail the question of the shifting prestige of humanities and science subjects, respectively. Here he took the Barthes-Picard Affair, a controversy which signalled the arrival of structuralism in the postwar French intellectual field, as exemplifying shifts in the forms and sources of academic prestige.

The Barthes-Picard Affair

Published sixteen years after the events of May 1968, *Homo Academicus* set itself the task of analysing 'the foundations and forms of power in the arts and social science faculties on the eve of 1968' (1984, p. 48 [p. 32]). Bourdieu sought to understand the events as the result of a series of structural transformations in the university field from the early 1960s onwards. For Bourdieu, the Barthes-Picard Affair had a particular significance as a 'microcosm' of these structural transformations and as a precursor of the political fault-lines along which different groups of lecturers and professors in 1968 would divide.

The Barthes-Picard Affair, a polemic between two literary critics, Roland Barthes and Raymond Picard, was provoked by two articles Barthes published in *Modern Language Notes* and the *Times Literary Supplement*. In these articles, he identified two conflicting schools of French literary criticism, 'university criticism', with its roots in a positivist tradition of literary history whose principles had first been formulated by Gustave Lanson early in the century, and 'the new criticism', which included figures such as Barthes himself, Lucien Goldmann, Jean Starobinski, and Gaston Bachelard, and which drew on structural linguistics, Marxism, psychology, and psychoanalysis to claim a scientific status for its findings (Barthes 1963a). These articles provoked an angry response from Picard, himself steeped in the tradition of Lanson, whose pamphlet *Nouvelle critique ou nouvelle imposture?* (1965) took issue with the assumptions behind the 'new criticism', concentrating specifically on Barthes's psychoanalytic and structuralist readings of Racinian tragedy in *Sur Racine* (1963).

The quarrel was widely reported in the French press, which divided along predictable lines, the right-wing *Le Figaro* defending Picard, whilst more left-wing or modernising magazines such as *Le Nouvel observateur* and *L'Express* lined up behind Barthes. As Barthes pointed out, what really seemed to have provoked Picard's ire was that French universities should have been criticised in a foreign publication and that the 'new criticism' should employ a language so at odds with the tenets of good taste. The honour of a national institution, a certain uncritical notion of the transparency of the French language, *'la clarté française'*, and an equally uncritical notion of 'taste' were all under assault from a discourse which drew on Marxism and psychoanalysis to emphasise the material bases of culture (Barthes 1966).

Bourdieu had offered his first tentative analysis of the Affair in the 1966 essay 'Intellectual Field and Creative Project' ('Champ intellectuel et projet créateur'). The importance of this early essay lay in the fact that it was the first occasion on which Bourdieu employed the term 'field' in its wider sense. Where in his analyses of the 'field of possibles' open to the Algerian sub-proletariat or the 'field of the photographable' which offered itself to different sections of the French population the term 'field' referred merely to an internalised set of 'objective possibilities', in 'Champ intellectuel ... ' the term took on the sense of a structured space of differential relations. A 'field' was analogous to a magnetic force-field with its poles of attraction and repulsion; it constituted a structured space of relations in which the positions of individuals or schools of thought were defined in terms of their differential relationship with other participants in the field. New entrants into the field were necessarily situated within this network of competing positions; the positions they took up were the result not of free choice but of the meeting between the field and their habitus, a socially inculcated structure of dispositions which ensured their investment in what was at stake in the field (Bourdieu 1966f).

Bourdieu has since argued that this uncritical investment in the stakes of the field, the 'illusio', generates a 'sense' or 'feel for the game', a 'practical' or implicit sense of which actions are likely to reap rewards, not a conscious search for prestige but a series of 'strategies' objectively orchestrated towards preserving or accumulating capital within the field. Further, it is this shared and unquestioned investment in the stakes of 'the game' which ensures that the most vehement disputes between participants in the field may in fact contribute to that field's reproduction. In the case of Barthes and Picard, the 'illusio' was both thinkers' unquestioned belief in the intrinsic value of studying a classic literary figure such as Racine. Thus Bourdieu argued that behind their apparent dispute lay a certain 'complicity', 'the *consensus* in *dissensus* which forms the unity of the intellectual field' (1966f, p. 902). The conflicting 'strategies' of the two adversaries reflected their struggle to impose their particular critical approach to Racine as legitimate and hence a source of further intellectual capital.

In 'Champ intellectuel ... ', Bourdieu limited his remarks to hinting at the 'complicity' between Barthes and Picard. In *Homo Academicus*, he offered a more detailed analysis of the Affair, placing it within the context of the structural changes affecting the university field of the 1960s and explaining the different 'strategies' adopted by Barthes, Picard and their supporters as responses to those changes. Further, by the time of *Homo Academicus*, Bourdieu had adopted a method of multivariate analysis known as 'correspondence analysis' as a means of plotting the co-ordinates of the intellectual field. Correspondence analysis is a means of first accumulating a set of data defining a given sample of institutions or individuals into 'profiles' of their attributes. In Bourdieu's case, this was a matter of gathering a mass of biographical data concerning the principal individuals in Parisian higher education establishments, their social origin, their cultural capital, their social capital, their membership of administrative committees, the institutions in which they were employed, and so on. Each individual's 'profile' is then plotted graphically so that individuals with similar characteristics will appear close to one another, in 'clusters', whilst those with different or opposing characteristics will be distant, occupying a different space or clustering around an opposing pole of the field.

Appropriating and adapting a terminology first coined by the philosopher of science Gaston Bachelard, Bourdieu argued that it was correspondence analysis which allowed him to undertake an 'epistemological break' between his own personal, anecdotal knowledge of the French intellectual field and a genuinely 'scientific' knowledge of that field. The individuals whose positions he plotted were no longer 'empirical individuals' but 'constructed individuals'; through a process of 'theoretical construction' they had become objects of science; they were located within a field of differential 'relations' and as such should be understood not as fixed substances but as entities defined 'relationally' or 'differentially' in terms of their position within a field of conflicting relations. Moreover, by locating or 'objectifying' his own position within the field, Bourdieu sought to work through and transcend the limitations of his partial point of view, his personal interests or anecdotal experience, to achieve a truly 'objective', 'scientific' grasp of the phenomena he described (1984, pp. 11–52 [pp. 1–35]).[6]

Using correspondence analysis, then, Bourdieu first plotted the French academic field of the 1960s in its entirety. The analysis revealed certain clearly distinguishable groups. The most significant distinction, Bourdieu argued, was that between those intellectuals with 'temporal' and those with 'spiritual' power. On the one hand, there were intellectuals with roots in the Catholic bourgeoisie who tended to occupy senior administrative positions, have right-wing political views, work in the law or medical faculties and 'compensate' for their lack of prestige in strictly intellectual terms with decorations for public service and influence over temporal power. Opposed to these were left-of-centre intellectuals, perhaps Jewish or agnostic, who concentrated on research, on the accumulation of strictly intellectual rather than social, political or administrative capital. This second group were most numerous in the science faculties (1984, pp. 55–96 [pp. 36–72]). Situated at the mid-point between these two poles, the 'sub-field' of the faculties of arts and social sciences reproduced this opposition in microcosm and as such represented 'the privileged vantage-point for observing the struggle between the two forms of university power', namely 'temporal' or 'worldly' power and 'spiritual' or 'scientific' power (p. 99 [p. 73]).

Figure 1 shows the results of Bourdieu's correspondence analysis of the 'sub-field' of the faculties of arts and social sciences. In order to emphasise the point that he was involved in a 'scientific' analysis of the field's 'objective' logic, rather than a series of ad hominem attacks, Bourdieu initially omitted the names of the individual intellectuals concerned, merely marking their various institutional affiliations and other characteristics. Subsequently, in an appendix he added both to the English translation and to the second French edition of 1988, Bourdieu decided to help his readers by adding the names he had previously left blank (see figure 2). On both graphics, the horizontal axis can be seen as measuring the forms of capital or prestige possessed by the individuals in question, whilst the vertical axis measures the amount of prestige or capital. Thus, those situated above the horizontal axis are older, established figures possessing more prestige than those situated below it, whilst those situated to the right of the vertical axis have greater institutional or temporal power within the university establishment than those situated to the left of the vertical axis, who possess greater intellectual prestige or capital.

In the bottom-left corner of the sub-field of arts and social sciences, therefore, were clustered those individuals, including Barthes, Michel Foucault, Jacques Derrida and Bourdieu himself, who were as yet marginal figures in the intellectual field. Typically employed in the Sixth Section of the École pratique des hautes études en sciences sociales (EPHE), their activities were oriented towards innovation and research which drew on the 'newer' social sciences such as sociology, psychoanalysis and structural linguistics. These individuals tended to side with Barthes during the Affair and were more sympathetic to the demands of the students in May 1968. In the bottom right-hand corner were clustered individuals and institutions, exemplified by Picard at the Sorbonne, whose efforts were directed towards the reproduction of legitimate culture and who pursued 'older' disciplines such as philology, classics or French literature. These individuals tended to side with Picard and sought to defend the universities against the criticisms of student radicals in May.

According to Bourdieu, the Barthes-Picard Affair was a product of struggles over which forms and sources of academic prestige should predominate at a time when the traditional hierarchy of academic disciplines was undergoing a dramatic inversion. The old hierarchy with philosophy and the humanities at its peak was being overturned by the increasing prestige accorded science. The field was at a critical point: 'a critical moment of the historical process which tends to subordinate the citadel of literary culture to scientific culture, which used to be subordinate' (p. 160 [p. 122]). Barthes's attempt to apply 'scientific' methods to the study of literature, through semiotics and psychoanalysis, constituted 'a reconversion strategy', an attempt to 'accumulate the benefits of science and the prestige of philosophy or literature', by 'reconverting' literary into scientific capital or rather accumulating the capital inherent to both at this transitional stage (pp. 154–5 [p. 117]). Bourdieu argued that behind the veneer of scientificity, Barthes remained allied

Academie Française Larousse 1968

>5 translations

Order of Merit

Honorary doctorate

Paris 16 or Neuilly

Institute Born before 1900

AXIS 2 4.30%

Father industrial management or ownership

Father large tradesman or merchant

Preparatory class in other provincial lycée

CNRS as second post

ANCIENT HISTORY

Oriental languages as second post

EPHE 4th or 5th section as second post

COLLEGE DE FRANCE

Father engineer Preparatory class in other Paris lycée

Father professor in higher education

Paris 13-14 Father craftsman

AXIS 1 5.17%

Television

EPHE 6th section as second post

Secondary education at lycée Condorcet or Carnot

Legion of Honour

Born 1900-4

CLASSICAL PHILOLOGY AND LITERATURE

Father primary schoolteacher

Normalien

Published in 'Idées' or 'Points' paperback series Private secondary school in Paris

Editorial committee

Secondary school lycée Louis-le-Grand

CNRS commission

2-6 citations (Citation Index) preparatory class in lycée Louis-le-Grand

No children

Council for Higher Education

Palmes Académiques

EPHE 4th, 5th sections PHILOSOPHY

Born large provincial city

Supported Faceliére in May 1968

Universities consultative committee

MODERN LITERATURE AND PHILOLOGY

Director of research team

Born abroad

1-5 translations

MEDIEVAL AND CONTEMPORARY HISTORY

Father officer

Born Central France

Born in Midi-Pyrénées region Who's Who

2 children

Born in Paris

Born Paris or Paris region 3 children

Agrégé

Other Paris suburbs or provinces or abroad

Born in Provence or Mediterranean France

Born in north of France

0-1 citations (Citation Index)

Father secondary schoolteacher

Preparatory class in lycée Henri IV

LANGUAGES

ENS board of examiners

Married Foreign or provincial lycée

Born eastern France

Born in central or eastern France

SORBONNE

Born in Paris suburbs

Order of Merit

Father tradesman

Unmarried

Father liberal profession father painter born medium-sized town

Honorary doctorate

Paris 16 or Neuilly

Born 1920-4

Supported Giscard d'Estaing

GEOGRAPHY

Born in S.W. France

4 children

Secondary school lycée J. de Sailly

no preparatory class

Born in Western France

Published in 'Que Sais-Je'? paperback series

Member of ministerial cabinet or Plan commission

5 or more children

Public or private secondary school

Father senior or middle executive

Born 1915-19 Paris 4, 15

ENS as second post

Paris 5, 6

Born 8, 17

Paris Born 1910-14

Secondary school private provincial Paris suburbs 78, 92

Born in other provinces

Paris 7

Born 1905-9

No translations

FACULTY OF NANTERRE

One child

Born 1920-4

Secondary school lycée Henry IV

Other Paris district

≥7 citations (Citation Index)

SOCIAL SCIENCE Institut des Sciences Politiques as second post

Father landowner

Secondary school lycée St. Louis

Writes in Nouvel Observateur

Supported Mitterrand

EPHE 6th section

Born in 1925 or after

Father farmworker, working class or white-collar worker

Preparatory class in major provincial lycée

Agrégation board of examiners

Figure 1 Faculty of Arts and Social Sciences – Properties

Source: Editions de minuit

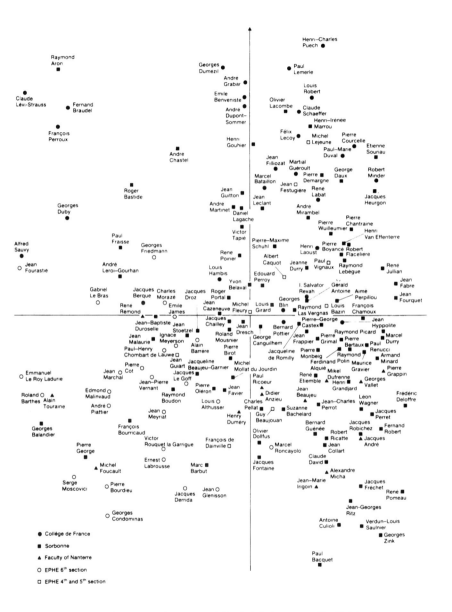

Figure 2 Faculty of Arts and Social Sciences – Individuals

Source: Editions de minuit

to a much older conception of the critic as creative genius: 'a creator able to re-create the work through an interpretation itself instituted as literary work and thus situated beyond the true and false' (p. 154 [p. 117]). Behind Barthes's critique of the Sorbonne's '*lansonisme*', Bourdieu saw the spectre of a classical assault on the 'reductivism' of positivist social science: 'this unending struggle between the "reductive materialism" of the social sciences, here incarnated in a perfect caricature, is henceforth achieved in the name of a science which, in the guise of semiology, or even structural anthropology, claims to be capable of reconciling the requirements of scientific rigour with the society elegance of authorial criticism' (p. 155 [p. 118]).

Thus Bourdieu could argue that Barthes and Picard were 'complicit adversaries [*des adversaires complices*]'; despite their differences, both remained committed to maintaining the prestige of literary study per se. Moreover, Barthes's attempt to adapt literary study to the demands of a new academic hierarchy, with the natural sciences at its peak, was merely one of a series of contemporary and analogous 'reconversion strategies' undertaken by a generation of 'consecrated heretics', such as Foucault, Deleuze, and Derrida, who occupied similar positions in the French intellectual field. Appropriating certain methodologies from social science, they had all sought to historicise or deconstruct philosophical concepts, but they had done so in such a way as to retain both their own prestige as 'master thinkers' and the traditional prestige of philosophy. Repeating an argument he had first advanced in *Distinction* (1979, pp. 578–83 [pp. 494–8]), Bourdieu maintained that, for example, Derrida's deconstructive readings of Kantian aesthetics in 'Economimesis' (1975) and *The Truth in Painting* (*La Vérité en peinture* 1978), were playing 'a double game'; apparently challenging the foundations of philosophical thinking, these deconstructive readings remained within the terms of philosophy as such. Unlike Bourdieu himself, whose reading of Kantian aesthetics revealed the latter's 'objective' social function, 'Derrida knows how to suspend "deconstruction" just in time to prevent it tipping over into a sociological analysis bound to be perceived as a vulgar "sociologistic reduction", and thus avoids deconstructing himself *qua* philosopher' (1984, p. 304 [p. xxiv]).

Further, Bourdieu suggested that Barthes, Foucault, Derrida and Deleuze were implicated in what he termed, the 'blurring of the differences between the field of restricted production', in which academic work was produced for and judged by an audience of one's academic peers, and 'the field of general production', in which a work's merit was judged by its marketability, its capacity to make an impression in the journalistic field (p. 157 [p. 120]). Barthes's challenge to Picard had involved appealing to a market composed of the increasing number of students, of newly recruited junior lecturers, and of graduates now working on the edges of the intellectual and the economic fields, in advertising, the media, and public relations, a market whose demands were fashioned 'by journalism with intellectual pretensions' (pp. 156–7 [pp. 119–20]). Whilst the Sixth Section of the École pratique could encourage innovation in research, it also encouraged its staff to seek recognition outside the 'autonomous' university field in compensation for their relative lack of recognition within that field; it thus constituted, 'the weak point in the university field's resistance to the intrusion of journalistic criteria and values' (p. 148 [p. 112]).

Thus Bourdieu argued that 'the true principle' behind the Barthes-Picard Affair, and by extension behind the emergence of structuralist and poststructuralist theories in 1960s France, was not to be found 'in the actual content of their respective declarations, which are simply rationalised retranslations of the oppositions between the posts held, between literary studies and social science, the Sorbonne and the École des hautes études, etc.' (p. 151 [p. 115]). Rather, this profusion of new theoretical approaches in philosophy, literary criticism, and the social sciences was to be understood as a strategic adaptation to changing forms and sources of academic prestige.

There is no doubt that Bourdieu's concept of 'field' and the use he made of correspondence analysis to plot that field's co-ordinates were powerful heuristic tools, revealing patterns of affiliation and opposition that might otherwise have remained obscure, whilst going a long way to explaining the resonance that the ideas of Barthes and his peers enjoyed amongst the broader French public. However, showing that Barthes's work was well-suited to exploit the new forms and sources of academic prestige does not necessarily prove that his work was either reducible to or motivated by the attempt to accumulate those new forms of academic prestige. There were two potential dangers implicit in Bourdieu's analysis. The first was to risk conflating questions belonging to the sociology of knowledge, the historical or institutional conditions under which certain forms of thought emerged, with questions of epistemology, the inherent validity of those forms of thought or knowledge. Whilst these two domains are related, the nature of that relation cannot a priori be assumed to be direct or straightforward and the value or significance of an intellectual movement is not merely reducible to the particular historical circumstances in which it emerged. The second risk was that of functionalism, the risk, that is to say, of mistaking outcomes for causes. To demonstrate that Barthes's innovations in literary study enabled him to exploit the new forms and sources of academic prestige is not to prove that such a beneficial outcome was the cause of those innovations, it may simply have been their fortuitous side-effect. Bourdieu's insistence that Barthes's position reflected not a conscious search for prestige but a 'strategy', a pre-reflexive investment in the stakes of the game, does not detract from the functionalist tendency of his analysis.

As Serge Doubrovsky pointed out in his study of the Barthes-Picard Affair, the fact that Barthes's work revealed a desire to 'reunite, through a thorny synthesis, the rhetorical value of literature and the truth claims of science' (1967, p. 10), did not prevent that work from having ramifications which extended far beyond questions of intellectual capital and were bound up with contemporary disputes over pedagogy and the outdated institutions of French higher education (p. xiv). This was a point which Bourdieu conceded in the preface he wrote for the English translation of *Homo Academicus*. Having previously implied that the 'true principle' behind the work of Foucault, Derrida, Deleuze and Barthes was a strategic reconversion to the increasing dominance of the sciences, Bourdieu maintained here that their works were, 'much more than more or less successful reconversions of the philosophical enterprise'; they partook of a more general critique of the institutions of higher education, 'against the archaic nature of their contents and their pedagogical structures', and were 'in harmony with the movements which pulsated through the ethical and political avant-garde of the student world' in the years around 1968 (1984, pp. 304–6 [pp. xxiv–xxv]).

This introduced a puzzling paradox; if the works of these thinkers were indeed 'much more' than simple reconversion strategies, if they fed into the protests and criticisms of French universities around May 1968, it was difficult to see how those who produced them could be considered 'consecrated heretics' whose struggles had merely contributed to the reproduction of the field. This implied that far from being 'the true principle' behind the struggles in the French university field of the 1960s, the need to adapt to the new forms and sources of academic prestige formed only one of a range of different determining principles. Alternatively, Bourdieu was arguing that the events of 1968 were themselves merely a strategic adaptation to the shifting balance of power within French higher education, an adaptation whose forms and wider significance were anticipated by the Barthes-Picard Affair and which hence left the system fundamentally unchanged.

The Events of May 1968

Bourdieu's analysis of the events of May 1968 in *Homo Academicus* picked up where *Reproduction* had left off, attributing a key role to the changes in the morphology of the student and lecturer populations.

Indeed, he argued that it was such morphological changes which, on a general level, 'introduced' historical change into fields whose semi-autonomy otherwise ensured their reproduction:

> The morphological changes here ... are the medium whereby history, which mechanisms of reproduction tend to resist, invests the fields, which are *open spaces* obliged to draw from outside the resources necessary for their functioning, and are thereby liable to become the locus of that collision of independent causal series which creates the event, that is, the quintessentially historical. (p. 49 [p. 33])

More specifically, Bourdieu argued that the sudden need for a mass of new lecturers to deal with the rapid expansion in student numbers had significantly changed the criteria of university recruitment. Traditional subjects such as philosophy or French literature had an existing 'reserve' of personnel, in the form of *agrégés* either working in or destined for secondary education, upon which they could draw. Newer disciplines such as sociology, which were not taught in the lycées and thus had no *agrégés*, had to employ personnel who did not have the same history of acculturation into the customs of French higher education. Thus, Bourdieu argued, where the newly recruited *agrégés* were predisposed to accept the rigid structures of French academe, its slow temporal rhythms, the long wait for promotion, the new recruits into social science departments were not so disposed. Their accelerated initial promotion into higher education had raised expectations of further advancement which had little objective chance of being met. This *décalage* or disjuncture between subjective aspirations and objective chances led to a growing sense of resentment and critique of the university authorities.

This sense of resentment was mirrored amongst the students. Bourdieu argued that with the massive increase in student numbers had come a 'generalised downclassing', a significant reduction in the value of a university qualification. A university degree was no longer a guarantee of future success and this was particularly true of disciplines such as sociology or psychology which had no well-defined career path and did not lead to employment in secondary education. Furthermore, these new disciplines tended to serve as 'refuge' for bourgeois students who had failed to gain entry into more prestigious degree courses and were 'endowed with expectations strongly maladjusted in relation to their chances of social success' (p. 222 [p. 170]). The effects of this 'downclassing' were not, Bourdieu maintained, limited to the universities alone. For, as the level of formal qualification demanded for every job rose, so adolescents with qualifications that would have previously guaranteed them a clerical post found themselves reduced to working on the factory floor. These young discontented workers, he argued, were to play a key role in the strikes of May.

Thus, a whole generation of young assistant lecturers [*assistants*] and lecturers [*maîtres-assistants*], of students and of young workers shared similar frustrations and occupied homologous, dominated positions in their respective fields. Similarly, those graduates who had moved into new jobs in the domain of media, cinema, advertising and marketing, 'the new agents of symbolic manipulation', felt an equivalent sense of frustration provoked by 'the opposition between their own representation of their task as intellectual creation in its own right and the bureaucratic constraints to which they must bend their activity' (pp. 228–9 [p. 175]).

The coincidence of these events in different fields, all endowed with their own relative autonomy, resulted in 'an effect of synchronisation' which provoked a more general crisis. This objective crisis was fuelled by various critical discourses whose combined effect was to introduce a 'rupture' or 'suspension' in the investment of groups or individuals in the continued smooth functioning of their respective fields:

> If the crisis goes hand in hand with criticism, it is because it introduces a break in duration, because it suspends the ordinary order of succession and the ordinary experience of time as presence in an already present future; in overthrowing in reality or in representation the structure of objective chances (of profit, of social success, etc.) to which behaviour reputed reasonable is spontaneously adjusted ... , the crisis tends to undo that sense of placing, both as 'knowing one's place' and as knowing how to place sound investments, which is inseparable from the sense of realities and possibilities which we call sensible. It is the *critical moment* when, breaking with the ordinary experience of time as simple reproduction of a past or of a future inscribed in the past, all things become possible (at least apparently). (p. 236 [p. 182])

Where, in describing the 'peasant experience of time' in Algeria, Bourdieu had turned to Husserl's conception of an internalised structure of 'retentions' and 'protentions', an implicit, 'practical', or 'doxic' sense of what was to come, *l'à-venir*, which precluded a reflexive return on to the potentially arbitrary nature of the social order, here he argued that that 'doxic' order had broken down; the morphological changes affecting French universities had provoked a 'suspension' or 'rupture', 'a veritable practical *epoche*' in the apparent self-evidence of the 'doxic order' (p. 305 [p. xxv]). The temporal rhythms of doxa, 'the ordinary experience of time as presence in an already present future', in which the structure of expectations internalised into the habitus was in accord with the objective chances of those expectations being realised, had broken down. As a result, time opened up into a kind of festive time, 'free, festive leisure time', 'a public time' in which 'everyone can believe that the processes of reproduction are suspended for a moment and that all futures are possible for all people' (pp. 237–40 [pp. 183–5]).

For Bourdieu, the crisis of May 1968 signalled the end of 'the cyclical time of simple reproduction' typical of 'traditional' societies, and this breakdown in the time of 'simple reproduction' within the universities was analogous to that he had analysed in his work on Algeria and the Béarn. He compared the need for the universities to change their criteria of recruitment to Halbwachs's analysis of French women's changing criteria for suitable husbands following the First World War (p. 183n.8 [p. 310n.8]). 'The crisis of succession', occasioned by the recruitment of an army of new *assistants* and *maîtres-assistants*, was equivalent to the crisis of succession he had described in the Béarn. Lecturers and professors who attempted to retain the prestige they had enjoyed at a previous moment in the system's history were like 'the elder sons of the "great" farming families in the Béarn region who were condemned to bachelorhood in the 1950s by their concern to avoid misalliance during a period of crisis on the matrimonial market' (p. 167 [p. 127]). The shock these same lecturers expressed when confronted with rebellious students reminded Bourdieu of old Kabyle peasants: 'Just like old Kabyle peasants speaking of the heretical methods of cultivation practised by the young, they can only express their stupefaction in the face of the *incredible*, the world upside-down' (p. 238 [p. 183]).

However, just as in his analysis of the erosion of the 'traditional' structures of Algerian society, so in his account of the universities in 1968, Bourdieu maintained that this breakdown in the 'cycle of simple reproduction' was insufficient to ensure that the participants in the events of May had an 'objective' understanding of the logic of the system against which they were rebelling. Echoing criticisms of the student movement he had first made in *The Inheritors*, Bourdieu argued that 'the movement unleashed by the aristocratic revolt of the students of bourgeois origin' had few chances of uncovering the role of the universities in the legitimation and reproduction of class divisions (p. 240 [p. 186]).[7] The alliances struck up between 'downclassed' students and workers were 'more or less fantasised' (pp. 214–15 [p. 164]), and the calls of the student leaders for students and workers to unite were often rooted in 'a magical denial of the factors causing the malaise' in education and society as a whole (p. 241 [p. 186]). Bourdieu argued that the teaching union most directly involved in the events, the Syndicat national de l'enseignement supérieur (SNESup), had an equally limited understanding of what was really at stake (pp. 241–2 [pp. 186–7]).

For Bourdieu, the 'most important and durable effect of the crisis' was 'the symbolic revolution as profound transformation of styles and thought of life, and, more particularly, of the whole symbolic dimension of everyday existence'. This 'symbolic' revolution had transformed 'the view which agents normally have of the symbolism of social relations, and especially the hierarchies, highlighting the otherwise strongly repressed political dimension of the most ordinary symbolic practices' (p. 250 [p. 193]). However, in 'La Production de l'idéologie dominante' (The Production of the Dominant Ideology) (1976), Bourdieu suggested that this 'symbolic' revolution, the demands for greater social, cultural and sexual freedom, had rapidly been recuperated into the dominant discourses of consumerism (1976, p. 44n.9; p. 55). In *Acts of Resistance* (1998), he argued that 'the reaction of retrospective panic provoked by the crisis of 1968' had paved the way for the hegemony of neo-liberalism at the expense of an earlier tradition of left-wing radicalism (p. 15 [p. 7]). It is in the light of Bourdieu's rather ambivalent verdict on the events of May that his reading of the theoretical innovations introduced by Barthes, Foucault, Deleuze and Derrida might be understood. His assertion that such innovations were both 'really' reconversion strategies and much more than that could be seen as a reflection of the essentially contradictory nature of both the intellectual shifts and the socio-political movements in French universities in the 1960s, movements which both challenged the status quo but were ultimately to be recuperated within a modified system.

The events of 1968 were a high point for the French New Left and this also goes a long way towards explaining Bourdieu's ambivalence to those events. As we have seen over the last two chapters, Bourdieu had expended considerable energy criticising what he saw as the sociologically flawed and politically compromised nature of the French New Left's analyses, whilst attempting to offer a more 'scientific', 'objective' account of developments in French culture and society in their place. In the case of education, this involved drawing heavily on the classical sociological tradition, appropriating and adapting Weber's concept of legitimation, a Marxist notion of class antagonism, and Durkheim's notions of reproduction, relative autonomy, 'downclassing' or '*déclassement*', and morphological change. Bourdieu was to examine the fallout of May 1968, the way that 'symbolic revolution' was recuperated by forces on the right of French politics, in more detail in both *Distinction* and *The State Nobility*. However, before turning to a discussion of those two studies, it will be necessary to return to the chronological development of Bourdieu's thought, to examine the 'theory of practice' which he had sketched in his three works of Kabyle anthropology between 1972 and 1980, 'the theory of practice' through which these later analyses of French education and culture would be interpreted.

For reasons of historical chronology, this chapter has discussed *Homo Academicus* out of sequence, as it were, introducing concepts such as 'field' and 'strategy', as well as Bourdieu's debt to the epistemology of Gaston Bachelard, before the development of such concepts has been examined in any real detail. By turning to an analysis of the 'theory of practice' Bourdieu elaborated in his anthropological studies of Kabylia, it will be possible to gain a much clearer understanding of the theoretical framework within which he sought to pursue and refine the analyses of French culture, class and education he had initiated in his work of the 1960s.

Notes to Chapter 3

1. For a more detailed analysis of the similarities and differences between the Bourdieusian and Althusserian theories of education, see Snyders (1976).
2. Claire Duchen's analysis of the effects of the postwar entry of women into French higher education and the workplace confirms Bourdieu's points about both the nature of women's expectations and the devalorisation of the disciplines and occupations to which they gained access (see Duchen 1994, pp. 150–1; p. 164).
3. Baudelot had originally been Bourdieu's student, contributing a chapter to *Academic Discourse*, before he broke with Bourdieu to join the Althusserians.
4. Numerous commentators have noted the extent to which Bourdieu's work influenced the nature of student demands in the years leading up to 1968 (see Aron 1968; Schnapp and Vidal-Naquet 1969, p. 17; Capdevielle and Mouriaux 1988, p. 211).
5. Aron and Bourdieu had fallen out over Bourdieu's qualified support for the students and workers in May 1968.
6. For a useful introduction to Bachelard's epistemological theories, see Bachelard (1971).
7. Bourdieu's conviction that the students had not really grasped the objective logic of the educational system against which they were protesting had also been made clear in a petition he published under the aegis of the CSE during the May events. Whilst expressing his sympathy for the student movement as a whole, he expressed reservations about the limitations of their understanding of what was going on in the universities, reminding them that: 'any questioning of the universities which does not centre on their function in eliminating the lower classes [*les classes populaires*], hence on the education system's function in social conservation, is necessarily imaginary' (in Schnapp and Vidal-Naquet 1969, pp. 695–6).

CHAPTER 4

Returning to Kabylia

In a footnote to *Reproduction*, Bourdieu stated that his 'theory of pedagogic action' was 'grounded in a theory of the relations between objective structures, the habitus and practice', which would 'be set out more fully in a forthcoming book' (1970, p. 9n.1 [p. xiii n.1]). The book in question was published two years later and drew on fieldwork Bourdieu had conducted during the Algerian War. Entitled *Esquisse d'une théorie de la pratique* (1972), it took the form of three anthropological studies of Kabylia followed by a sustained reflection on the political, ethical, and epistemological implications of anthropological study. Five years later, Bourdieu revised and entirely rearranged *Esquisse ...* in preparation for its English translation as *Outline of a Theory of Practice* (1977). Three years later, he returned to the subject of Kabylia once more, revising his initial analyses one further time before publishing them as *The Logic of Practice* (*Le Sens pratique* 1980)

The frequency with which Bourdieu has returned to data that were by 1972 already ten years old suggests that these works of Kabyle anthropology occupy an extremely important position within his broader intellectual project. Indeed, the three works, *Esquisse ... Outline ...* and *The Logic of Practice* can be situated at a transitional stage in Bourdieu's career; they represent a kind of theoretical hiatus during which he was able to elaborate the concepts of habitus, practice, and strategy into a full-blown 'theory of practice', before applying that theory to the analyses of class, culture and education he was to offer in later works such as *Distinction* or *The State Nobility*. Clearly, then, if these later works on class, culture and education are to be properly understood, it will be necessary first to undertake a critical analysis both of the 'theory of practice' Bourdieu elaborated in his Kabyle work and of the place of that Kabyle work within his wider *oeuvre*, that is to say of the relationship between the anthropological and sociological aspects of his work. This chapter will focus on Bourdieu's theory of practice, leaving discussion of the relationship between anthropology and sociology in his work to the next chapter.

The importance of Bourdieu's Kabyle works is not limited to their role in the development of his thought. For these works have exerted a considerable influence over the field of anthropology internationally. Indeed, Bourdieu's attention to the ethical, political and epistemological implications of anthropological study has led to his works being caught up in a series of debates sparked by what has been variously termed the 'reflexive' or 'postmodern' turn in anthropology. The notion is a somewhat imprecise one but, broadly speaking, it refers to the increasing willingness of anthropologists to reflect on their discipline's relationship to the history of Western colonialism and hence to question the ways in which anthropologists have typically classified and represented the anthropological 'Other', the 'primitive' or 'exotic' societies and peoples they have conventionally studied. Frequently, this questioning of anthropology's ethical and epistemological foundations has drawn on the work of French poststructuralists such as Jacques Derrida, Michel Foucault or Gilles Deleuze, who, despite their many differences, have all shown a concern to challenge the ethnocentric premises of the Western philosophical tradition in the name of a greater sensitivity to questions of 'difference' or 'alterity', to the problematics of representing the Other.

The fact that Bourdieu's works of Kabyle anthropology emerged out of a critical encounter with structuralism and were marked by a sustained reflection on the problematics of anthropological representation has led some critics to situate Bourdieu's work firmly within this postmodernist or poststructuralist movement (Lash 1990, pp. 237–65). For example, in their seminal collection, *Writing Culture: the poetics and politics of ethnography* (1986), James Clifford and George Marcus identify Bourdieu's 'theory of practice' as one key influence on the reflexive turn in anthropology, placing it alongside the work of contemporary French thinkers such as Foucault and Derrida (1986, p.3). Bourdieu himself, however, has responded with hostility to any suggestion of an affinity with this broad current of thinkers. Indeed, he has attacked the *Writing Culture* collection for its 'falsely radical denunciation of ethnographic writing as "poetics and politics"', which 'opens the door to a form of thinly veiled nihilistic relativism ... that stands as the polar opposite to a truly reflexive social science' (Bourdieu 1992, p. 52 [p. 72]). A number of Bourdieu's more favourable critics have picked up on his criticisms of 'postmodern' anthropology to suggest his work offers a more scientific, historically grounded, but nonetheless reflexively vigilant alternative to postmodernism's perceived 'relativism' or 'nihilism' (Thomas 1994, pp. 8–9; pp. 58–60; Calhoun 1996). A range of other critics, however, have suggested the opposite, arguing that Bourdieu's portrayal of Kabylia is marked above all by its stasis and ahistoricism, qualities they identify as characteristic of precisely the kind of ethnocentrism Bourdieu claimed to have worked through (Lacoste-Dujardin 1976; Herzfeld 1987, pp. 7–8; pp. 83–4; Caillé 1994, pp. 163–4; Reed-Danahay 1995; Free 1996).

Adjudicating between these diametrically opposed interpretations will require focusing on issues of epistemology, history and representation, as well as on the historical agency ascribed to the subjects of Bourdieu's studies, the Kabyles themselves. Whilst concurring with Bourdieu that his commitment to scientificity did indeed distinguish his work from that of the postmodernists, this chapter will seek to assess his claims that this allowed him to offer a scientific, historically specific, non-ethnocentric representation of the Kabyles. It will start by focusing on his attempts to elaborate a truly reflexive epistemology, show how this leads to the development of concepts such as 'practice', 'habitus', and 'strategy', all the while examining to what extent these concepts live up to the claims that have been made for them both on the theoretical level and from a more empirical, historical standpoint.

Epistemology: 'The Scholastic Point of View'

Bourdieu's frequent returns to his Kabyle material can be explained, in part at least, by the extent to which anthropology provided him with a striking paradigm of the pitfalls attendant on any intellectual discourse about the social world. The very tangible distance between Western anthropologists and the non-Western societies they studied was merely, he suggested, a more extreme example of the distance that separated all detached, leisured bourgeois intellectuals from the social and cultural phenomena they sought to interpret. Failure to think through the distortions inherent in this distance was, moreover, manifest in the failings of the two opposing intellectual currents against which Bourdieu would define his approach in his studies of Kabylia, namely Sartre's existential phenomenology and Claude Lévi-Strauss's structural anthropology. According to Bourdieu, the opposition between Sartre's existentialism and Lévi-Strauss's structuralism was merely the most recent manifestation of a false dichotomy between subjectivism and objectivism which had long impoverished sociological and anthropological thinking.

Given that he had been initiated into the French intellectual field in the 1950s and 1960s, it was understandable that Bourdieu should see that field as being marked by a dichotomy between subjectivism and objectivism, existentialism and structuralism. For the postwar period in France had seen the initial dominance of the existential and phenomenological philosophies of Sartre and Merleau-Ponty, with their emphasis on the dramas of the individual *subject*, being challenged and displaced, from the late 1950s onwards, by the hegemony of structuralist modes of thought, which proclaimed 'the death of the subject' and focused on *objective* linguistic and cultural structures. Bourdieu set out to identify and transcend the limitations of these opposed schools of thought, locating such limitations in the failure of either school adequately to think through the distortions inherent in the distance separating intellectuals from the objects of their studies.

Structural anthropology, for example, was a clear case of a social theory that had failed to think through the distortions inherent in the distance that separated observer from observed. In reducing complex social practices to a closed system of structural laws intelligible only to the impartial observer, structuralists forgot that what appeared to the detached intellectual as a reified set of formal rules, constituted for native participants a set of practical problems to be negotiated with relatively unpredictable outcomes. Sartrean existentialism, meanwhile, committed an analogous error, albeit in reverse: it attributed to every agent a freedom in fact contingent upon the privileged status of the intellectual. In each case, intellectuals were guilty of unconsciously importing their relationship to the object of study into their analysis of that object: 'just as objectivism universalises the theorist's relation to the object of science, so subjectivism universalises the experience that the subject of theoretical discourse has of himself as subject' (Bourdieu 1980, p. 77 [pp. 45–6]). In neither case had intellectuals reflected upon the social, historical, and material determinants of that relationship. As Bourdieu put it in *Esquisse* ... :

> We might suggest that the experience of a social world upon which one can act, as if almost by magic, by using signs (words or money), that is to say through the mediation of someone else's labour, does not predispose one to perceive the social world as a place of *necessity* and enjoys a certain affinity with a theory of action as either the mechanical execution of a model or the pure expression of free choice, this depending upon whether one is thinking more of oneself or of others. (1972, p. 195)

Moreover, Bourdieu argued, the intellectual attitude that led to this sort of theoretical distortion was not limited to the domain of anthropology alone. By a series of analogies, he suggested that this kind of failing was characteristic of intellectual discourse about a range of other social phenomena, from aesthetics to education. Thus, just as Sartre attributed a universal value to freedoms that were contingent upon his status as a bourgeois intellectual, so Kantian aesthetics elevated certain historically and socially determined principles of taste to the status of universals, positing, as Bourdieu put it in *Distinction*, a '*homo aestheticus* ... as the universal subject of aesthetic experience' (1979, p. 577 [p. 493]).

Structuralism too had an affinity with Kantian aesthetics; both were types of formalism. For just as the Kantian aesthete appreciates a work of art on the level of pure form, as 'a purposefulness without purpose', an object of beauty in and of itself, so structuralism understood linguistic or cultural phenomena as a kind of formal game, a set of logical operations to be decoded: 'as an autonomous, self-sufficient object, that is a purposefulness without purpose – without any other purpose, at any rate, than that of being interpreted, like a work of art' (1980, p. 53 [p. 31]). The condition of possibility of both these formalisms, Bourdieu argued, was a certain bourgeois '*aisance*', a certain distance from the constraints of material need which allowed the aesthete to suspend all interest in the practical utility of the work of art and ensured that the interests of structural linguists or anthropologists were not at stake in the social practices they described. Detached from the issues at stake in practice, structural anthropologists were free to view the society under study in purely theoretical terms, as a spectacle laid out for their contemplation, a code to be deciphered, a reified structure of fixed rules or disembodied forces beyond the ken or control of individual agents.

Playing on the etymology of the term 'scholastic', from the Greek *skholè*, meaning 'leisure', Bourdieu argued that such typically 'intellectualist' failings were the product of what he termed the 'scholastic point of view', of a particular disposition inculcated in and rewarded by the academic world: 'a product of the scholastic situation, in the strong sense of a situation of *skholè*, of *otium*, of *inactivity*' (1980, p. 53 [p. 31]). Thus, four of the main strands of Bourdieu's thought coalesced around the single concept of the 'scholastic point of view': Bourdieu had already attempted to trace the genesis and effects of this 'scholasticism' in his work on education, whilst he was to develop his critique of its application to the fields of aesthetics, in *Distinction*, linguistics, in *Language and Symbolic Power* (*Ce que parler veut dire* 1982), and anthropology, in his Kabyle work.[1]

Bourdieu (1994a, p. 221 [p. 127]) has indicated that the phrase 'scholastic point of view' derived from J.L. Austin's *Sense and Sensibilia* (1962, pp. 3–4) and this has led several critics to suggest a significant affinity between his thought and the Anglo-American philosophical tradition, a suggestion encouraged by Bourdieu himself (see Shusterman, ed. 1999). However, Bourdieu's own declarations of an affinity to the Anglo-American tradition should be understood as a primarily strategic move, an attempt to distance himself from Continental Philosophy and hence from the post-structuralist and postmodernist philosophies with which his work has frequently and mistakenly been associated by various commentators. This strategic move should not be allowed to obscure the far more decisive influence of the work of both Merleau-Ponty and Bachelard on Bourdieu's critique of 'the scholastic point of view'. Merleau-Ponty's criticisms of the pitfalls attendant on the opposing doctrines of 'empiricism' and 'intellectualism' in *Phenomenology of Perception* (1945) and his consequent emphasis on the inherently embodied nature of human 'practice' anticipated in important respects both Bourdieu's critique of the 'scholastic point of view' and his own 'theory of practice'. The particular form taken by that critique, meanwhile, owed much to the work of Bachelard.

Bourdieu acknowledged this debt to Bachelard when he explained that the frequent revisions and re-workings of his Kabyle work represented an instance of 'epistemological vigilance', that constant process of rectification and self-critique in which earlier errors were worked through and dialectically overcome. As he put it in *Esquisse* ... : 'the history of successive errors forms an integral part of a body of scientific knowledge whose fundamental gains were only achieved by reflecting on those errors' (1972, p. 129n.1). As Bourdieu had indicated in *The Craft of Sociology* (*Le Métier de sociologue* 1968), this vision of (social) scientific knowledge as being an open-ended process in which prior errors were to be worked through and dialectically overcome, owed much to Bachelard:

> As the whole *oeuvre* of Gaston Bachelard shows, epistemology differs from abstract methodology inasmuch as it strives to grasp the logic of error in order to construct the logic of the discovery of truth as a polemic against error and as an endeavour to subject the approximated truths of science and the methods it uses to methodical, permanent rectification. (1968, p. 9 [p. 3])

The Craft of Sociology, a collaborative study in which Bourdieu and his co-authors attempted to provide secure epistemological groundings for sociological research, can be seen as providing the epistemological principles which Bourdieu attempted to put into practice in his Kabyle work, rather as Durkheim's *The Rules of Sociological Method* (1894) formalised the principles he had put into practice in *The Division of Labour in Society* (1893). The book took the form of a lengthy theoretical introduction followed by a series of 'illustrative texts' drawn from a characteristically eclectic range of thinkers, Marx, Weber, Durkheim, Bachelard, Erwin Panofsky, Charles Darwin and Georges Canguilhem, amongst others. The precedence given Bachelard's work was evident both in the Introduction and in the titles of the various sub-sections under which the 'illustrative texts' were grouped; 'The Break', 'Constructing the Object', 'Applied Rationalism' all referred to key Bachelardian concepts.

In *La Formation de l'esprit scientifique* (The Production of the Scientific Mind) (1938), Bachelard had argued that scientific knowledge could only progress by 'a psychoanalysis of objective knowledge', by constantly reflecting upon and thinking through the irrationalisms, outdated problematics, errors, and 'epistemological obstacles' contained in every scientist's 'epistemological unconscious'. Bourdieu's determination to think through the unconscious *social* determinants of sociological discourse, of 'the scholastic point of view', can be seen as an effort to transfer this method from the natural sciences to the social sciences, to, as he put it in *The Craft of Sociology*, 'extend the "psychoanalysis of the scientific mind" by an analysis of the social conditions in which sociological works are produced' (1968, pp. 9–10 [p. 3]).

Phenomenology and structuralism thus represented for Bourdieu two 'epistemological obstacles', 'an epistemological pair', in the Bachelardian sense, blocking the further advance of social scientific knowledge. Working through this 'epistemological pair' did not involve 'breaking' with structuralism and phenomenology in any straightforward sense. Rather it was a question of locating and working through the blind spots implicit in each approach, whilst integrating their respective insights in a dialectical synthesis which would 'transcend the antagonism which opposes these two modes of knowledge, while preserving the gains of each of them' (1980, p. 43 [p. 25]). This classically dialectical movement also had a precedent in Bachelard's work. For example, in *Le Nouvel esprit scientifique* (The New Scientific Mind) (1934), when Bachelard explained the advances secured by non-Euclidean geometry over what had gone before, he emphasised that this advance was not won simply by denying the insights of Euclidean geometry but rather by locating Euclidean geometry's specific realm of application within a more advanced synthetic general geometry (1934, p. 12). It was precisely this procedure that Bourdieu would effect with regard to phenomenology and structuralism, each theory being not denied but rather integrated and re-positioned within an overarching, synthetic 'theory of practice'.

For Bourdieu, a truly scientific sociology involved a double movement. Firstly, it was necessary to undertake 'an epistemological break', to achieve a moment of objectification in which the hidden logic of social processes was laid bare. When Ferdinand de Saussure first claimed the autonomy of the linguistic system, '*langue*', from its individual moments of actualisation, '*parole*', he had undertaken one such epistemological break. However, Bourdieu argued, this epistemological break had also concealed a social break which brought with it the dangers of a theoretical distortion. In appropriating Saussure's model for anthropological study, structural anthropologists had overlooked the significance of this social break, leaving an unanalysed 'epistemological unconscious' at the heart of their work (1980, p. 51 [p. 30]).

This 'epistemological unconscious' could, however, be thought through by means of a second epistemological break, a reflexive return on to the first break's social, historical, and economic conditions of possibility. This allowed social scientists to acknowledge the potential for theoretical distortion inherent in their project and to grasp that level of 'practical knowledge' deployed by native agents which unreflected objectivism had necessarily excluded. Theoreticist or objectivist modes of thought relied on maintaining the opposition between theoretical and practical knowledge firmly in place; they could never transcend that opposition by integrating those two modes of knowledge in a dialectical synthesis:

> the triumphalism of theoretical reason is paid for in its inability, and this from the very beginning, to move beyond simply recording the duality of the paths to knowledge, the path of appearances and the path of truth, *doxa* and *episteme*, common sense and science, and its incapacity to win for science the truth of what science is constructed *against*. (1980, pp. 61–2 [p. 36, trans. modified])

'From the very beginning', that is, from Plato's *Republic* onwards, science or *episteme* has been constructed against common sense or *doxa*.[2] Yet for Bourdieu, a truly scientific account of the social world must include the element of *doxa*, the realm of practical knowledge or common sense, so as 'to win for science the truth of what science is constructed against'. This turn to the doxic had, Bourdieu insisted, 'nothing to do with the phenomenological recovery of the lived experience of practice' (1972, p. 156). Although Bourdieu's critique of objectivism clearly owed much to phenomenology, it was not a matter, as Husserl had argued in his *Crisis of European Sciences and Transcendental Phenomenology* (1954), of overcoming the excessive rationalism of the Western philosophical tradition by simply overturning the opposition between *episteme* and *doxa* and making the 'disparaged *doxa* ... a foundation for science' (1954, pp. 155–6). If *doxa* constituted an integral part of all social phenomena, it did not, for all that, contain their objective logic, which could only be revealed by means of the kind of objectifying epistemological break undertaken by structuralism. Thus, scientific knowledge of the social world required the dialectical synthesis of both theoretical and practical modes of thought, of *episteme* and *doxa*, of the objective knowledge secured by structuralism and the subjective knowledge secured by phenomenology. As Bourdieu later put it:

> it is not sufficient to break with ordinary common sense, or with scholarly common sense in its ordinary form. We must break with

the instruments of rupture which cancel out the very experience against which they have been constructed. This must be done to construct more complete models which *encompass both the primary naiveté and the objective truth that this naiveté conceals.* (1992, p. 221 [p. 251], my emphasis).

This synthesis of *episteme* and *doxa* would thus conserve the gains of each whilst transcending their limitations, producing what Bourdieu variously termed 'praxeological knowledge' (1972, p. 164), 'a sort of third order knowledge' (1977a, p. 4), or more simply 'a higher objectivity' (1980a, p. 32 [p. 17]).

Bourdieu was clearly attempting, here, to overcome the traditional opposition between theory and practice. Indeed, he went so far as to argue that the activity of the social scientist needs itself to be understood as constituting a '*practice*', a 'theoretical' or 'sociological practice', whose epistemological principles have been incorporated, as a set of 'practical precepts', into a 'scientific habitus ... that functions in a practical state', forming a transposable scientific disposition applicable to a range of experimental contexts (1968, pp. 10–11 [pp. 3–5]; 1992, pp. 192–4 [pp. 221–4]). In figuring sociology as a 'craft' rather than a purely intellectual activity, the title of *The Craft of Sociology* further emphasised this attempt to overcome the opposition between theory and practice, whilst alluding to Martin Heidegger's aphorism '*Denken ist Handwerk*' and recalling those sections of *Being and Time* (1927) in which Heidegger compared '*Dasein*' to the relationship between a craftsman and his tools, a relationship irreducible to objectivist analysis.[3]

However, Bourdieu's apparent willingness to dispense with the distinction between theory and practice seemed inconsistent with his own descriptions of practice as corresponding to a pre-predicative or pre-reflexive form of consciousness, an immediate investment in the apparent self-evidence of the social world, which precluded, by definition, a reflexive return on to its objective logic. For example, in *The Logic of Practice*, he defined 'the truth of practice as a blindness to its own truth', going on to state that 'practice excludes any reflexive return [*la pratique exclut le retour sur soi*]' (1980, pp. 153–4 [pp. 91–2, trans. modified]). To argue that the precepts of a scientific sociology should become internalised in 'a practical state' into the habitual, pre-reflexive structures of a 'scientific habitus' would seem, by Bourdieu's own definition, to place such precepts beyond the grasp of that reflexive return necessary for Bachelard's imperative for 'epistemological vigilance' to be respected. The notion that the principles of a scientific sociology should become a set of *habitual*, 'quasi "unconscious"' dispositions seemed equally inimical to both the spirit and letter of Bachelard's epistemology. As Bachelard put it: 'In many ways, scientific *method* is the antithesis of *habit* and it is the gnoseological error of formalism to want to render method *mechanical*. Our consciousness of method must remain vigilant' (1949, p. 25).

Bourdieu attempted to escape this impasse by arguing that epistemological vigilance was one of the internalised dispositions which made up the scientific habitus (1968, pp. 11–12 [p. 5]). However, this merely reintroduced the old opposition between theory and practice in the form of a new distinction between the scientific habitus, which allowed for a reflexive return on to its own logic, and the habitus possessed by 'ordinary' social agents, which excluded any such reflexive return. Ultimately, then, Bourdieu would insist on the need to keep the opposition between practical and theoretical knowledge firmly in place:

> If one must objectivise the schemata of practical sense, it is not for the purpose of proving that sociology can offer only one point of view on the world among many, neither more nor less scientific than any other, but to wrench scientific reason from the embrace of practical reason, to prevent the latter from contaminating the former, to avoid treating as an instrument of knowledge what ought to be an object of knowledge, that is everything that constitutes the practical sense of the social world, the presuppositions, the schemata of perception and understanding that give the lived world its structure. (1992, pp. 215–16 [p. 247])

As we have seen, according to Bourdieu to 'objectivise the schemata of practical sense' demands that sociologists and anthropologists achieve an epistemological break with the realm of pre-reflexive or practical immediacy. However reflexively undertaken, the ability to achieve such a break was, by Bourdieu's own definition, the preserve of the detached bourgeois intellectual. Logically, then, objective knowledge about the Kabyle or any other social world remained the exclusive privilege of the detached intellectual observer. As George Marcus (1998, pp. 194–6) has pointed out, it is this problematic that surely explains the extent of Bourdieu's hostility to the 'nihilism' and 'relativism' of that 'postmodern turn' in anthropology exemplified by the *Writing Culture* collection. For the contributors to this collection have challenged the notion that the detached anthropological observer has privileged access to objective truths about the social world which are not accessible to native participants. They emphasise instead that anthropological knowledge is necessarily a narrative construct, the product of an ongoing 'dialogical' encounter between observer and observed. For all Bourdieu's attentiveness to the distortions inherent in the status of the detached observer, this was a step he was not prepared to countenance since it would have involved abandoning any claim to a privileged measure of objectivity and scientificity.

There was, thus, at the heart of Bourdieu's epistemology, an unresolved tension between a stated desire to undermine the oppositions between theory and practice, subjectivism and objectivism, and a need to reassert those very oppositions in order to maintain his claim to scientificity. Indeed, Bourdieu's critique of the 'scholastic point of view' rested on the a priori assumption that there was a qualitative difference between the kind of knowledge of the social world accessible to detached intellectuals, on the one hand, and native agents, on the other. Whilst Bourdieu's intention was clearly to challenge the somewhat grandiose theorising characteristic of structural anthropology, this distinction between intellectualist and practical modes of thought cut both ways. For, if the intellectual's experience of the world was intrinsically mediated, distant and objectifying, this seemed to imply that native experience was, by contrast, irrevocably immediate, pre-reflexive, and subjective.

As Bourdieu put it in *The Craft of Sociology*, 'the sine qua non for the constitution of sociological science' was 'the principle of non-consciousness', the a priori assumption that native agents remained unconscious as to the objective logic of their own practices (1968, p. 38 [p. 16]). This 'principle of non-consciousness', of course, had a direct impact on questions of agency, of the extent to which the Kabyles themselves were considered able to know and transform their social world through willed praxis. Indeed, closer analysis of the concepts of 'habitus', 'practice', and 'strategy' will reveal that they too were marked by an unresolved tension between an emphasis on their creative, improvisational qualities and Bourdieu's insistence that such improvisations were always ultimately determined by an 'objective logic', the imperatives of social reproduction, which the Kabyles themselves 'misrecognised'.

'The Kabyle House' and The Poetics of Space

Of the three anthropological studies which opened *Esquisse* ... , perhaps the best known is 'The Kabyle House or the World Reversed' ('La Maison kabyle ou le monde renversé'). Originally written in 1963–64, it had first been published in 1970 in a volume of tributes to Lévi-Strauss on his sixtieth birthday. Described by Bourdieu as 'the last work I wrote as a blissful structuralist', a reworking of the essay, which reflected his critique of structural anthropology, appeared in all three of *Esquisse* ... *Outline* ... and *The Logic of Practice*. For this reason and because the Kabyle house represented for Bourdieu a microcosm of the entire mythico-ritual structure of Kabylia, the essay provides a good way to examine the theory of practice Bourdieu elaborated in critical reaction to both structuralism and phenomenology.

At first glance, 'The Kabyle House or the World Reversed' reads like a classical example of structural anthropology, 'a most impressive *tour de force* of structuralist analysis', as Richard Jenkins (1992, p. 32) puts it. According to Bourdieu, the Kabyle social universe was structured by a series of primary spatial oppositions between inside and outside, east and west, which were overlaid with a complex network of arbitrary and differential symbolic meanings that reflected and reinforced the gender divisions of the Kabyle world. The interior of the house was a dark, humid space signifying feminine values of nurture, domesticity and reserved respectability, or *'h'urma'*. Outside was the domain of men. Men went out into the fields to work, to meet and talk with other men. Male honour, or *'nif'*, was associated with this movement out into the world, with risk-taking and self-projection. When they went out to work in the morning, men left by the east door, walking out into the rising sun, leaving the dark domestic space behind them to the west. Just as the outside signified masculine values of light, openness and honesty when compared with the dark, feminised domestic interior, so the east was associated with masculine values and the west with feminine.

The interior of the Kabyle house reproduced these oppositions in microcosm. Thus, although when viewed from the outside the house embodied purely feminine values, once inside it was subdivided into masculine and feminine spaces. However, within the house the poles were reversed. Where external space was divided between a masculine eastern pole and a feminine western one, within the house it was the western wall that was associated with masculinity since it was lit by the light of the rising sun, whilst the eastern wall was in shade. The orientation of values was thus overturned within the house, hence 'the world reversed' of the essay's title, and the threshold, as the point where these opposing sets of values met and were inverted, took on a powerful symbolic function.

Bourdieu's analysis was clearly greatly indebted to structuralism. The network of cultural meanings which overdetermined the spaces and objects of the Kabyle world formed a structure of binary oppositions and homologies in a manner seemingly typical of structuralism. However, unlike a classically structuralist analysis, the meaning of each signifying element, the objects or spaces which formed the pole of each binary opposition, could not be deciphered merely by reference to a transcendent code irreducible to its individual moments of actualisation. The meaning of the domestic space, for example, was not solely dependent upon its position within an overarching structure of difference, but altered depending upon the gender, perspective and bodily position of the individual apprehending it. Thus, the domestic space might signify purely feminine values from the outside, or both masculine and feminine values from within. Further, for a woman the domestic interior signified positive values of domestic respectability, whilst for a man it signified a realm of essentially negative, feminised values. As Bourdieu explained: 'one or other of the two systems of oppositions that define the house, either in its internal organisation or in its relationship with the external world, is brought to the fore depending on whether the house is considered from the male or the female point of view' (1970a, p. 753).

The locus of meaning had thus shifted from the system of binary oppositions laid bare by the detached intellectual observer to the movements and perspectives of the Kabyles themselves. Meanings were generated not at the level of a purely theoretical, disembodied structure of difference but by the Kabyles in their everyday actions and movements, in their 'practice', as Bourdieu was to put it in *Esquisse ... Outline ...* and *The Logic of Practice*.

This emphasis on changing perspectives and on the embodiment of symbolic meaning suggested Bourdieu was drawing as much on the phenomenological as the structuralist tradition. In *Le Déracinement* (1964), written at the same time as 'The Kabyle House', he had sought to distinguish between the de-humanising, geometrical space of the 'resettlement centres' and the traditional space of the Kabyle house, rich in symbolic meaning and collective memory. Significantly, Bourdieu had quoted the following passage from Bachelard's phenomenological study of domestic space, *The Poetics of Space* (1957):

> Over and above our memories, the house we were born in is physically inscribed in us. It is a group of organic habits. ... In short, the house we were born in has engraved within us the hierarchy of the various functions of inhabiting. ... The word habit is too worn a word to express this passionate liaison of our bodies, which do not forget, with an unforgettable house. (Bachelard 1957, pp. 14–15, quoted in Bourdieu 1964, p. 152n.1).

Clearly, in 'The Kabyle House', Bourdieu was attempting to convey precisely this sense of domestic space as 'a group of organic habits'. Indeed, the oppositions between inside and outside, dark and light, female and male, which Bourdieu found structuring the Kabyle house seemed to echo the oppositions between inside and outside, between the dark cellar and the light, airy attic, which Bachelard had identified in *The Poetics of Space*. Moreover, Bachelard had laid special emphasis on the symbolic function of the threshold, quoting the following fragment of a poem by Michel Barrault:

> I find myself defining the threshold
> As being the geometrical place
> Of the comings and goings
> In my Father's House (in Bachelard 1957, p. 223)

Commenting on this fragment, Bachelard had written: 'the poet does not shrink before reversals of dovetailings [*le renversement des emboîtements*], ... he actually experiences the reversal of dimensions [*le renversement des dimensions*], or inversion of the perspective of inside and outside' (p. 225). It could be argued that the very title of 'La Maison kabyle ou le monde *renversé*' was derived from Bachelard's phenomenological study of domestic space. Indeed, in all three of *Esquisse ... Outline ...* and *The Logic of Practice*, Bourdieu quoted this same fragment, praising the poet's ability to 'go straight to the principle of mythopoeic practice', to grasp the significance of the threshold in its symbolic function as the site of the passage of socially marked trajectories, of the union, separation, and inversion of previously opposing principles, east and west, masculine and feminine, dominant and dominated (Bourdieu 1980, p. 158 [p. 94]).

If 'The Kabyle House' contained elements drawn from both the structuralist and the phenomenological traditions, the revisions to which Bourdieu subjected this essay in his later works of anthropology were less a matter of breaking with his earlier approach than of synthesising more completely the diverse elements on which it had drawn. Thus, from structuralism Bourdieu retained the notion of a socio-cultural universe structured by a set of binary oppositions which formed what he termed the 'dominant taxonomy'. However, he emphasised that these 'classificatory schemata' constituted *'practical* taxonomies'; they existed not as a set of theoretically coherent, abstract, disembodied rules but were incorporated into the 'hexis' and 'habitus' of the Kabyles, to generate a collective 'practice' which obeyed a 'practical' or 'fuzzy' logic not reducible to reified structural law.

The division between the sexes, which had been identified in 'The Kabyle House' as the fundamental socio-cultural opposition structuring Kabyle society, was thus described in *The Logic of Practice* as being rooted in the Kabyles' 'hexis' or bodily deportment rather than in a set of logical rules or norms, whether consciously or unconsciously obeyed:

> The opposition between male and female is realised in posture, in the gestures and movements of the body, in the form of the opposition between the straight and the bent, between firmness, uprightness and directness (a man faces forward, looking and striking directly at his adversary), and restraint, reserve, and flexibility. (1980, pp. 117–18 [p. 70])

Where the term 'hexis' referred to the style of deportment that resulted from the incorporation of a set of socially and culturally overdetermined oppositions, 'habitus' was employed in a more general sense to describe the whole series of bodily and cognitive dispositions which were both structured by past experience and structuring of future action. The habitus represented a durable and transposable structure of dispositions, an incorporated set of 'practical taxonomies', of 'classificatory schemata', of ways of seeing and doing in the world, 'a vision and division of the social world':

> The *habitus*, a product of history, produces individual and collective practices – more history – in accordance with the schemes generated by history. It ensures the active presence of past experiences, which, deposited in each organism in the form of schemes of perception, thought and action, tend to guarantee the 'correctness' of practices and their constancy over time, more reliably than all formal rules and explicit norms. (1980, p. 91 [p. 54])

This account of the habitus as the product of a historical process of bodily incorporation stood in clear contrast to Lévi-Strauss's structural analyses of social norms and rules. For Lévi-Strauss had posited a direct correspondence between the binary structure of linguistic or cultural systems and certain unchanging, fundamental structures of the human unconscious. Paul Ricoeur had famously argued that in Lévi-Strauss's thought the unconscious functioned as a kind of transcendental matrix structuring all experience and denying any constitutive role to the rational agent; structural anthropology was 'a Kantianism without transcendental subject' (1969, p. 55). Bourdieu reiterated this charge of idealism:

> Beneath its air of radical materialism, this philosophy of nature is a philosophy of mind which amounts to a form of idealism. Asserting the universality and eternity of the logical categories that govern the 'unconscious activity of the mind', it ignores the dialectic of social structures and structured, structuring dispositions through which schemes of thought are formed and transformed. (1980, p. 69 [p. 41])

If Bourdieu's emphasis on the dialectical relationship between habitus and social structure had a strong Marxist resonance, his insistence on the embodied nature of symbolic meaning recalled the phenomenological critique of Kantian idealism. In *Phenomenology of Perception*, Merleau-Ponty had argued that the nature of experience could not be explained by reference to a set of a priori categories, but rather lay in the relationship between living body and lived world:

> The Kantian subject posits a world, but, in order to be able to assert a truth, the actual subject must in the first place have a world or be in the world, that is carry around about itself a system of meanings whose correspondences, relationships and

involvements need not be made explicit in order to be made use of. (1945, p. 129, trans. modified)

This notion of the subject-in-the-world carrying around a series of meanings and correspondences which need not be explicitly posited to be practically deployed is very close to Bourdieu's own description of the structures of meaning attached to the divisions of the Kabyle social universe, which were realised in the Kabyles' practice rather than existing as a set of logical, a priori rules.

According to Bourdieu, Kabylia possessed few, if any, objectified systems of social regulation and no specific institutional forms of education (1977a, p. 17). Thus, legitimate forms of thought and action were inculcated by what he termed 'an implicit pedagogy', by the social imperatives conveyed through myth, proverb, ritual, and the most apparently banal parental injunction to 'sit up straight' or 'hold your knife in your right hand', which imposed a whole vision and division of the social world on to the somatic structures of practice.[4] Similarly, Bourdieu argued that the collective experience of living in and moving through a social universe structured by a series of cultural and symbolic meanings was enough to, 'impose the integration of bodily space with cosmic space and social space by applying the same categories ... both to the relationship between man and the natural world and to the complementary and opposed states and actions of the two sexes in the division of sexual labour and the sexual division of labour' (1980, p. 130 [p. 77]). Social imperatives were inculcated pre-thetically by a process of 'mimesis, a sort of symbolic gymnastics ..., from body to body, i.e. on the hither side of words or concepts' (1977a, p. 2).

Again, this account of 'incorporation' was greatly indebted to Merleau-Ponty, who had argued that motor skills such as dancing or typing were 'picked up' by the body in a way which a purely objectivist analysis of their constituent elements would be unable to grasp: 'As has often been said, it is the body which "catches" (*kapiert*) and "comprehends" movement. The acquisition of a habit is indeed the grasping of a significance, but it is the motor grasping of a motor significance' (1945, p. 143). Merleau-Ponty argued that the possession of such motor skills as typing or dancing did not reflect the deployment of a set of logical rules but rather a *practical knowledge* or '*praktognosia*':

Our bodily experience of movement is not a particular case of knowledge; it provides us with a way of access to the world and the object, with a '*praktognosia*' [practical knowledge], which has to be recognised as original and perhaps as primary. My body has its world, or understands its world, without having to pass through any 'representations', without subordinating itself to a 'symbolic' or 'objectifying function'. (pp. 140–1, trans. modified)

It was precisely this practical knowledge, which passed 'from practice to practice without passing through explanation and through consciousness', that Bourdieu found at the heart of the Kabyles' practice and habitus and which he argued objectivism was unable to grasp (1972, p. 190).

Having thus appropriated from structural anthropology the notion of a social and cultural world structured by a set of taxonomies, of binary oppositions and homologies, which could only be laid bare by effecting an epistemological break with the realm of first-hand experience, Bourdieu took from phenomenology the notion that such taxonomies were 'incorporated' into the kind of bodily disposition analysed by Merleau-Ponty to form part of what Husserl had identified as the 'practical' or 'pre-predicative' bases of experience.[5] Where Bourdieu's account differed from classical phenomenology was in its insistence on the material determinants of this process of incorporation. Husserl and Merleau-Ponty understood practice and incorporation from the point of view of, respectively, the transcendental Ego or the purely subjective experiences of an individual body. Bourdieu, however, saw the body as the site of the incorporation of a series of profoundly social and historical forces: 'the body as geometer, a "conductive body" run through, from head to foot, by the necessity of the social world' (1980, p. 245 [p. 145]).

For both Husserl and Merleau-Ponty, this embodied practical knowledge was endowed with an intentionality, the body acted on the world by being directed towards, by aiming at or tending towards a particular goal or *telos*, in what Merleau-Ponty termed *'une visée'* or *'une visée corporelle'*. In his Kabyle work, Bourdieu would reinterpret this intentionality in specifically sociological terms, placing it at the heart of his concepts of 'field', 'interest' and 'strategy'. In *Experience and Judgement* (1948), Husserl had described the ego's perception of a given object as consisting of a series of past apperceptions, retentions, which provoked a series of protentional 'anticipations', an 'aiming' or 'tending-towards', an 'interest' in accumulating knowledge about those aspects of the object as yet unknown, 'a progressive *plus ultra*'. This 'interest', Husserl argued, had 'nothing to do with a specific act of will', but rather was inherent in the ego's being-in-the-world, inherent in 'every act of the turning-toward of the ego, whether transitory or continuous, every act of the ego's being with (*inter-esse*)' (1948, pp. 80–6).

In Merleau-Ponty's work, this kind of intentionality became less a matter of cognition and apperception than of embodied practice. In a passage which anticipated Bourdieu's conception of the relationship between field, habitus and strategy, Merleau-Ponty pointed to the footballer's practical knowledge of the football field:

For the player in action the football field is not an 'object', that is, the ideal term which can give rise to an indefinite multiplicity of perspectival views and remain equivalent under its apparent transformations. It is pervaded with lines of force (the 'touch lines'; those which demarcate the 'penalty area') and articulated in sectors (for example, the 'openings' between the adversaries), which call for a certain mode of action and which initiate and guide the action as if the player were unaware of it. The field itself is not given to him, but present as the immanent term of his practical intentions; the player becomes one with it and feels the direction of the 'goal', for example, just as immediately as the vertical and horizontal planes of his own body. It would not be sufficient to say that consciousness inhabits this milieu. At this moment consciousness is nothing other than the dialectic of milieu and action. Each manoeuvre undertaken by the player modifies the character of the field and establishes in it new lines of force in which the action in turn unfolds and is accomplished, again altering the phenomenal field. (1949, pp. 168–9)

It was this implicit sense of where the limits and boundaries of certain practices lay, of which 'moves' were most likely to pay off, that Bourdieu saw as the key to the 'strategies' of the Kabyles, 'strategies' generated at the intersection between their habitus and the social field they inhabited. Applying Husserl's and Merleau-Ponty's insights to the domain of social interaction, Bourdieu attributed to the Kabyles an inherent 'tending towards' and 'interest' in the conservation or accumulation of 'symbolic capital', whether through gift exchange or judicious matrimonial strategies. This 'interest' reflected less a conscious striving for personal gain than a 'doxic' or pre-reflexive adherence to 'the *illusio* in the sense of *investment* in the game and its stakes, of *interest* for the game, of adherence to the presuppositions – *doxa* – of the game' (1980, p. 111 [p. 66]). Native participants in any social field deployed their 'practical sense', an internalised sense of which moves would prove most profitable; 'an almost bodily tending towards the world [*visée quasi corporelle du monde*] ... , a proleptic adjustment to the demands of a field, what, in the language of sport, is called the "feel for the game" as a "sense of placement" [*sens du placement*], art of "anticipation", etc.' (1980, p. 111 [p. 66]).[6]

Although not motivated by a conscious striving for power or prestige, Bourdieu argued, Kabyle practice was nonetheless objectively oriented to that end and as such was endowed with a kind of intentionality without conscious intention, a purposefulness without explicitly articulated purpose: 'It is because native membership in a field implies a feel for the game, the art of practically anticipating the forthcoming [*l'à-venir*] contained in the present, that everything that takes place in it seems *sensible*, objectively endowed with sense and objectively oriented in a judicious direction' (pp. 111–12 [p. 66, trans. modified]). It was this intentionality without conscious intention, this 'interest' in the accumulation of symbolic capital which, Bourdieu argued, formed the objective logic behind the Kabyles' various 'strategies'.

From a Rule to a Strategy

Bourdieu's most detailed account of the concept of 'strategy' can be found in his analyses of gift exchange and kinship. Once again, these analyses represented an attempt to overcome dialectically the opposition between phenomenology and structuralism. In the case of gift exchange, his analysis was explicitly positioned as an attempt to work through and transcend the limitations inherent in the phenomenological and structuralist accounts of this phenomenon offered by Marcel Mauss and Lévi-Strauss, respectively, offering 'a paradigmatic illustration of the theory of the relations between the three modes of theoretical knowledge, that is to say, the mode of phenomenological knowledge, in Mauss's analysis, the mode of objectivist knowledge, in Lévi-Strauss's analysis, and the praxeological analysis' (1972, p. 245n.8).

In his 'Introduction to the Work of Marcel Mauss' (1950), Lévi-Strauss had argued that Mauss's phenomenological analysis in *The Gift* (1923–24) had fallen short of grasping the objective logic of the gift, namely the structural law of exchange, the necessity to give and receive which determined all gift exchange (in Mauss 1950, pp. IX–LII). This analysis, Bourdieu argued, betrayed the failings typical of structuralism. By understanding gift exchange as an inevitable process in which the first gift was always reciprocated, Lévi-Strauss ignored the fact that the exchange took place over time and that this temporal aspect was intrinsic to the structure of the exchange. To give a gift in return immediately would cause insult since it would reveal the selfish intention behind the initial act of giving. Moreover, it would mean that the function of gift exchange, the exercise of power by the donor over the recipient through the latter's sense of obligation, could not be achieved: 'the functioning of gift exchange presupposes misrecognition of the truth of "the objective mechanism" of the exchange, the very truth that would be brutally revealed if the gift were immediately returned' (Bourdieu 1972, p. 223).

Thus, as Bourdieu put it, 'to abolish the interval between gifts is to abolish strategy' (ibid). The strategies employed by the Kabyles in gift exchange consisted precisely in playing off their implicit sense of how long the return of a gift could be deferred, whether it could simply be refused, what the implications in terms of loss or accumulation of symbolic capital would be of giving to which individuals, when, and how often. Gift exchanges thus had a characteristic rhythm and tempo, and to maximise the symbolic profits on offer required all the skill of a virtuoso: 'only the virtuoso who perfectly masters his "art of living" can play off all the resources offered him by the ambiguities and uncertainties of different behaviour and situations in order to produce actions suited to every occasion'. It was this virtuosity which defined what Bourdieu termed 'excellence', a 'practical mastery' of social norm and convention which was neither the result of free choice nor of mute submission to structural rule or law, but was '"the art" of *necessary improvisation* which defines excellence' (p. 226).

The participants in gift exchange were, thus, neither freely choosing when and to whom to give, nor were they the unconscious puppets of immutable structural law, rather they were involved in a dialectical process in which social and historical circumstance met the generative schemes of the habitus to produce a relatively unpredictable outcome. For the system to function, its participants had to possess a certain partial or practical knowledge of what was at stake and which 'moves' would prove profitable or detrimental. Yet, Bourdieu argued, they also had to remain ignorant of the objective logic of gift exchange, of the arbitrary nature of the social hierarchies it produced and reproduced. This combination of partial knowledge with a pre-reflexive investment in the stakes of the game, this structure of 'misrecognition/recognition' was at the heart of gift exchange: 'in order for the system to function, agents cannot be completely ignorant of the schema which organise their exchanges and whose logic is revealed by the anthropologist's mechanical model, yet at the same time they cannot know or recognise that logic' (p. 223).

Bourdieu's use of a series of musical metaphors – virtuosity, rhythm, tempo, improvisation, was clearly intended both to echo and subtly to inflect the complex orchestral motifs employed by Lévi-Strauss in the *Mythologiques* (1964–71) series. In *The Raw and the Cooked* (1964), for example, Lévi-Strauss had compared South American tribal myth and ritual to a classical orchestral score, arguing that 'the myth and the musical work are like conductors of an orchestra, whose audience becomes the silent performers' (1964, p. 17). In emphasising the improvisational nature of social practice, Bourdieu challenged the formal closure of this account, countering that the practices of the Kabyles were 'collectively orchestrated without being the product of the organising action of a conductor' (1980, p. 89 [p. 53]). Similarly, in his analysis of matrimonial exchange, Bourdieu emphasised that only those matches with major political implications, such as those which joined partners from different tribes or clans, demanded that official conventions be followed to the letter. Only in these exceptional cases would it be appropriate to explain social practice by analogy with a written musical score: 'because the stakes are important, the risks of failure so numerous and so great, one cannot rely on the regulated improvisation of orchestrated habitus and every action must become the performance of a musical score' (1972, p. 115).

In understanding social action as the execution of a musical score, Lévi-Strauss had overlooked the possibility that what native informants posited as the official rules regulating marriage might in fact constitute less a rule than, by turns, an ideal hardly ever realised, a duty of honour which could be broken in certain circumstances, or simply one possible 'move' amongst others. Bourdieu argued that parallel-cousin marriage, identified by Lévi-Strauss as the structural rule or norm behind kinship exchange, was empirically speaking relatively rare. The real dynamic behind such exchanges was not structural rule but strategy, the Kabyles' own practical knowledge and implicit sense of which matches were objectively possible, given historical circumstance and the status of their own family, and which would best serve to preserve or accumulate that family's symbolic capital. That native informants themselves frequently referred to parallel-cousin marriage as the rule regulating matrimonial exchange represented an instance of 'second order strategies which aim to conceal the strategies and interests pursued behind the appearance of obedience to the rule' (1972, p. 95).

The concept of strategy was, therefore, clearly intended to provide an account of social practice as the dynamic or dialectical interaction between habitus and social structure, internalised disposition and an objectively and historically determined set of possible alternatives. In this sense, it seemed to echo Marx's famous remark in *The Eighteenth Brumaire* that: 'Men make their own history ... but under circumstances ... given and transmitted from the past' (1852, p. 103). However, Bourdieu seemed ambiguous about the extent to which the strategies of the Kabyles were the result of a *conscious choice* amongst a historically determined range of alternatives, or simply the product of a *practical or implicit sense* of what was to be done. Equally, it was unclear whether the ability to undertake strategies with relatively unpredictable outcomes in gift and matrimonial exchange really amounted to a capacity on the Kabyles' part to make their own history or effect significant social change. For Bourdieu's replacement of Lévi-Strauss's vocabulary of structural rule and law with the notions of improvisation and strategy concealed the fact that these strategies were determined by an immanent, hypostatised law, namely that tendential law which apparently demanded that all strategies aim at the preservation or accumulation of symbolic capital. Moreover, Bourdieu maintained that the result of this tendential law was to orient Kabyle society towards the reproduction of its existing social structure:

> The practices observed in a social formation oriented to the simple reproduction of its own foundations ... can be analysed as the product of strategies (conscious or unconscious) through which individuals or groups aim to satisfy the material and symbolic interests which are associated with the possession of a material and symbolic inheritance and which hence tend to ensure the reproduction of that inheritance and, at the same time, of the social structure. (1972, p. 119)

According to this account, strategies were anything but transformative of social structure; social change could, apparently, only come from changes in material circumstance and not from the willed praxis of the Kabyles themselves. The habitus represented an internalised set of expectations, an implicit sense of what could or could not be achieved which reflected objective probability, attested to by past experience and internalised into a structure of durable and transposable dispositions. As long as objective conditions remained unchanged, Bourdieu argued, an almost perfect fit between objective probability and subjective expectation would ensure that every action had the appearance of a self-evident necessity. Questioning the logic of such actions would thus be precluded and the cycle of simple reproduction preserved:

> the practices that are engendered by the habitus and are governed by the past conditions of production of their generative principle are adapted in advance to the objective conditions *whenever the conditions in which the habitus functions have remained identical, or similar, to the conditions in which it was constituted.* Perfectly and immediately successful adjustment to the objective conditions provides the most complete illusion of finality, or – which amounts to the same thing – of self-regulating mechanism. (1980, p. 104 [p. 62], my emphasis)

By definition, then, this cycle of simple reproduction could only be broken by changes to the objective conditions which structured the habitus, and not vice versa.

Thus, whilst the concepts of habitus, practice and strategy were able to account for improvisation and relative unpredictability in a way structural anthropology could not, ultimately the ramifications of that improvisation and unpredictability remained strictly limited. As Michel de Certeau pointed out, Bourdieu's texts on Kabylia were characterised by a peculiar tension:

Scrupulously examining practices and their logic ... , the texts finally reduce them to a mystical reality, the *habitus*, which is to bring them under the law of reproduction. The subtle descriptions of ... Kabylian tactics suddenly give way to violently imposed truths, as if the complexity so lucidly examined required the brutal counterpoint of dogmatic reason. (de Certeau 1980, p. 59)

As de Certeau (1980, p. 80) noted, this tension was rooted in Bourdieu's epistemology, in his insistence that ultimately only the detached observer could grasp the objective, scientific logic of Kabyle practice. Only by attributing the Kabyles' strategies to an imperative for social reproduction of which they remained unconscious could Bourdieu maintain the opposition between his theory of their practices and those practices themselves on which his claim to scientificity ultimately depended.

History, Writing and the Body

Thus, even as Bourdieu sought to emphasise the improvisational and relatively unpredictable logic of Kabyle practice, so his account seemed to resurrect the very determinism he wished to counter. For all the apparently open-ended, unpredictable nature of the Kabyles' strategies, they ultimately remained subordinate to the imperatives of social reproduction, to the maintenance of the social system. Historical change, Bourdieu argued, could come not from contradictions, questions, or challenges thrown up from within the system itself, but only from the intrusion of external historical determinants:

It is out of the question that this elastic logic of overdetermined, 'fuzzy' relationships, protected from contradiction or error by its very weakness, might encounter within itself the obstacle or resistance capable of determining a reflexive return or questioning. *History can therefore only befall the system from the outside*, through the contradictions arising from synchronisation (favoured by writing) and from the systematising intention that it expresses and makes possible. (1980, p. 438n.59, [p. 316n.43, trans. modified, my emphasis])

The role Bourdieu attributed to writing, here, was significant. For he seemed to be opposing writing, as a medium of *re-presentation*, to the immediate self-presence of practice. In practice, social rules and conventions were realised through bodily actions, present only in the immediacy of their realisation. Thus practice excluded the possibility of a reflexive return on to its objective foundations:

Caught up in 'the matter in hand', totally present in the present and in the practical functions that it finds there in the form of objective potentialities, practice excludes any reflexive return onto itself (that is, onto the past). It is unaware of the principles that govern it and the possibilities they contain; it can only discover them by enacting them, unfolding them in time. (p. 154 [p. 92, trans. modified])

Writing, it appeared, could arrest this unmediated temporal flow of practice, inscribing social rules and conventions into a synchronic medium which allowed for reflection upon their potentially arbitrary nature and comparison with alternative social systems. Writing, Bourdieu argued, transformed the whole relationship to the body and to time, permitting reflexivity and freedom of action on a scale previously impossible:

What is learned by the body is not something one has, like a knowledge one can hold up and point to, but something that one is. This is particularly clear in societies without writing, where inherited knowledge can only survive in the incorporated state. It is never detached from the body that bears it and can be reconstituted only by means of a kind of gymnastics destined to evoke it ... , the body is thus constantly mingled with all the forms of knowledge which it reproduces and which never have the objectivity nor the freedom with respect to the body that objectification in writing ensures. (p. 123 [p. 73, trans. modified])

It was precisely this opposition between societies with and societies without writing that Derrida had sought to deconstruct in his reading of the work of Lévi-Strauss in *Of Grammatology* (1967). Derrida did not question the existence of pre-literate societies, nor did he deny that the spread of literacy or the invention of technologies of printing could have a significant effect on socio-economic and political development. Rather, he questioned the use of the absence of writing or literacy in certain societies as foundation for an ethnocentric opposition between the immediate or unmediated nature of experience in 'primitive' societies and the mediated, reflexive nature of experience in the 'modern' West. Derrida employed the term 'writing' in an extended sense to refer to any medium of re-presentation in which the apparently immediate self-presence of the spoken word was deferred or mediated. He argued that to imagine a state before the advent of any structure of mediation or deferral, the unmediated self-presence of the spoken word or, in Bourdieu's case, the temporal immediacy of embodied practice, was to construct a typically ethnocentric representation of 'primitive' societies as the site of some primal unity. If the advent of writing, from somewhere outside 'primitive' society, was seen to contaminate this primal unity, it was also seen as the source of all change, progress and history.

In opposing the mediated structures of writing to the immediacy of Kabyle practice and linking the advent of writing, from somewhere outside Kabylia, to historical progress, social change, and free agency, Bourdieu's analysis seemed to rest on precisely those ethnocentric oppositions which Derrida had sought to deconstruct in his readings of Lévi-Strauss. Moreover, these descriptions of practice as inherently unmediated or immediate seemed to contradict Bourdieu's own detailed analyses of the complex structures of Kabyle myth and ritual. For myth and ritual were clearly highly mediated structures of representation which appealed to cognitive rather than purely 'practical' or bodily dispositions. Bourdieu, however, distinguished between two senses of the French word '*représentation*', contrasting the more philosophical sense of the word, implying a reflexive grasp on the social world, with its use in the sense of a theatrical performance. Myth and ritual, he argued, constituted representations in the second sense of the term; they were 'practical representations', a tacit bodily relation to the world rather than an articulated structure of belief or cognition:

> These ritual manifestations are also representations – in the theatrical sense – performances, shows that stage and present the whole group, which is thus constituted as the spectator of a visible representation *which is not a representation of the natural and social world, a 'world-view' as we like to say, but a practical and tacit relationship to the things of the world.* (Bourdieu 1980, p. 184, [p. 108, my emphasis, trans. modified])

Ironically, in a conversation with the Kabyle poet and anthropologist Mouloud Mammeri published two years before *The Logic of Practice*, Bourdieu had warned against the ethnocentrism implicit in the assumption that an oral culture could not contain a reflection on the nature of the social world as conscious and profound as any written tradition:

> People cannot conceive that *oral* and *popular* poems might be the product of an intellectual effort, in their form as much as their content. People cannot accept that such poems might be composed in order to be declaimed in front of an audience of ordinary men and contain an esoteric meaning, which is thus destined to be meditated and commented upon. Of course, they exclude the possibility that a poem might be the product of a conscious effort to communicate *a level of figurative meaning* using the codified and objectified procedures characteristic of oral improvisation, such as iteration. (1978a, p. 51)

If this passage, with its insistence on the objectified and reflexive structures of the Kabyle oral tradition, seemed directly to contradict the oppositions between practice and writing which underpinned Bourdieu's analysis in his book-length studies of Kabylia, there were also good empirical historical reasons for questioning the adequacy of the description of Kabylia as a 'society without writing'. The fact that Kabylia had, since the seventh century, been an Islamic society meant that it possessed one clearly identifiable, highly codified, written, objectified set of moral and social precepts. Indeed, in the course of their conversation, Bourdieu and Mammeri emphasised the agonistic nature of the relationship between the oral tradition of the Kabyle bards and the written tradition of the Islamic marabouts, who, prior to the arrival of the French, were the representatives of State power, personified by the Turkish Dey (1978a, pp. 62–3). The arrival of the French in the nineteenth century had presumably rendered this *rapport de forces* more complex still.

However, although Bourdieu's book-length studies of Kabylia contained passing references to the presence of mosques and Koranic schools, whilst *Esquisse ...* told of Kabyle parents going to register their new-born children with the French authorities (1972, p. 38), he never explained what the relationship was between these three competing and contradictory traditions, the Kabyle, the Islamic and the French colonial. Nor did he explain how the existence of these codified and institutionalised forms of political and religious power could be squared with his claim that Kabylia was a 'society without writing', a society in which social imperatives were inculcated exclusively in practical form, a society characterised by 'the absence of political institutions endowed with an effective monopoly of legitimate violence' (1977a, p. 40).

It is precisely this puzzling absence of any discussion of the relationship between the customs and values of Kabyle society at the local level and 'encompassing bureaucratic and religious institutions' that leads Herzfeld (1987, p. 8) to qualify Bourdieu's analyses in *Outline ...* as 'Eurocentric'. Such 'Eurocentrism', Herzfeld argues, is rooted in the 'false distinction between *societies with codified laws* and *societies with a customary morality*' (p. 83). Reed-Danahay (1995), meanwhile, points to the absence of any discussion of Koranic or French colonial schools in Kabylia as evidence of Bourdieu's construction of an 'essentialised' Kabylia, 'a model of ... traditional society as a counterpart to Western society'. Thus, according to Herzfeld and Reed-Danahay, Bourdieu would appear to have reproduced the very ethnocentric tropes and oppositions which he had set out to avoid.

In sketching out his 'theory of practice', Bourdieu had sought to work through and transcend the conventional oppositions between objective and subjective, theoretical and practical modes of thought. There can be no doubt that Bourdieu's critique of 'objectivism' and the 'scholastic point of view' usefully highlighted some of the distortions typical of an 'intellectualist' viewpoint on the social world. Further, his consequent emphasis on 'practice' and 'strategy' revealed aspects of social practice which structural anthropology, in particular, tended to overlook. Nonetheless, Bourdieu's analyses of Kabylia did ultimately appear to rest on a series of binary oppositions between incorporated practice and objectified or written law, presence and representation, immediacy and mediation, societies without writing and societies with writing, oppositions which seemed to resurrect the very distinctions between subject and object, practice and theory that Bourdieu claimed to have transcended. For critics such as Herzfeld and Reed-Danahay, these oppositions are grounded in a single, fundamentally ethnocentric opposition between Kabylia and the West, between 'primitive' and 'modern' societies. However, as the next chapter will demonstrate, closer analysis of the relationship between the anthropological and sociological aspects of Bourdieu's work reveals that the situation is rather more complex and accusations of ethnocentrism are thus somewhat premature.

Notes to Chapter 4

1. For a more detailed discussion of the 'scholastic point of view', see Bourdieu (1994a, pp. 221–36 [pp. 127–40]; 1997, pp. 61–109).

2. See Book VII of Plato's *Republic* for the opposition between the realm of science or truth (*episteme*) and the realm of mere opinion, common sense, or appearance (*doxa*) (Plato 1930, vol.2, pp. 119–233).

3. In *Esquisse* ... , Bourdieu had quoted these very sections from Heidegger to illustrate his notion of the 'practical' or 'doxic' relation between the Kabyles and their social world (1972, p. 202; p. 213). These references to Heidegger were removed from both *Outline* ... and *The Logic of Practice* presumably because in the interim Bourdieu had published his first analyses of Heidegger's Nazism, a series of articles which first appeared in book form in German in 1976, before appearing in French as *L'Ontologie politique de Martin Heidegger* (1988), immediately following the controversy sparked by Victor Farias's *Heidegger et le nazisme* (1987).

4. In emphasising the importance of this kind of education in the formation of children's habitus and hexis, Bourdieu was also echoing Mauss's essay 'Les Techniques du corps' (in Mauss 1950, pp. 365–86).

5. For Husserl's emphasis on the 'pre-predicative' basis of experience see particularly the opening sections of *Experience and Judgement* (1948, pp. 11–101).

6. The expression '*sens du placement*' is almost untranslatable since it contains two specific meanings in French which cannot be encompassed

within a single English phrase. It refers firstly to the sense of placement of a sportsperson, the ability of a tennis player to place their body in anticipation of where they sense their opponent's ball will bounce and to place their return out of reach of their opponent. Secondly, the French term '*un placement*' means a financial investment, so to have a good '*sens du placement*' means to know which investments will prove most profitable. It is this dual sense of practical anticipation and symbolic profit that is implied in Bourdieu's usage.

CHAPTER 5

Anthropology and Sociology

In the spring of 1980, the year in which the French edition of *The Logic of Practice* was first published, Kabylia was shaken by a series of strikes, mass demonstrations, and violent anti-government protests. Generally referred to as 'the Berber Spring', the events of April 1980 were sparked by the decision of the ruling Islamic and Marxist FLN to ban a lecture given by Mouloud Mammeri at the University of Tizi-Ozou to promote a bilingual French-Kabyle collection of poetry, *Poèmes kabyles anciens* (1979). A violent reaction against merely the latest in a series of attempts by the FLN to extinguish any expression of a specifically Kabyle identity, the 'Berber Spring' represented, as numerous commentators have remarked, a significant challenge to the FLN's political and cultural hegemony, a challenge whose effects are still being played out today (Ouerdane 1990, pp. 183–98; Djender 1992; Colonna 1996, pp. 3–4).

'The Berber Spring' might be seen as a prime example of those 'local-national interactions' whose 'neglect' Reed-Danahay and Herzfeld have both noted in Bourdieu's works of Kabyle anthropology; it resulted from the tensions between national institutions, in the form of the university and of a ruling party trying to impose a uniform Arabic culture on all of Algeria's ethnic groups, the legacy of colonialism, in the form of a book of poems published in French by a Parisian publishing house, and 'traditional' Kabyle culture, in the poems themselves. Indeed, it was precisely this interpretation of events which Bourdieu offered in an article on the subject published in April 1980 in the French daily newspaper, *Libération*.

Bourdieu argued that the 'Berber Spring' could only be understood by taking into account the peculiarities of Kabylia's history under colonialism. He noted the importance of 'the Kabyle myth', an idealised vision of Kabyle society, which had been fostered by generations of French colonial ethnographers and administrators, according to which the Kabyles were racially distinct from and innately superior to Algeria's Arab populations. As Algeria's 'noble savages', the Kabyles had been seen as prime candidates for assimilation to French culture and civilisation.[1] It was 'the Kabyle myth' which had encouraged post-independence Algerian governments to view any expression of a distinct Berber identity as a form of neo-colonialism. This same 'myth' explained the relatively high number of colonial schools which had been situated in Kabylia, affording the Kabyles access to education and encouraging the history of labour migration between Kabylia and France. As Bourdieu explained, both factors had contributed to the Kabyles' historically high level of politicisation, as witnessed not only in the 'Berber Spring', but also in the predominant role they had played in the formation of the Algerian independence movement. Expanding on the conversation he had published with Mammeri in 1978, Bourdieu thus concluded that the events in Kabylia needed to be placed within the context of a longer history of tensions between 'indigenous' Kabyle culture, Islam, and French colonialism:

> Mouloud Mammeri has shown in great detail that the incessant transactions between the Berber heritage and the Islamic norm were the product of a quasi-conscious political game whose protagonists were the Koranic scholar (the *marabout*), the Berber sage (the *amousnaw*) and the Kabyle peasant. If the Berber tradition has been conserved, it is at the price of a sort of thousand-year-old effort to conserve through compromise and dialogue with the politically or symbolically dominant institutions. This defensive transaction continued in relations with the culture of the colonisers. (Bourdieu 1980c)

What was striking about Bourdieu's analysis of the situation in Kabylia in the *Libération* article was the extent to which it contrasted with the analyses of the same society he had offered in his book-length studies of Kabyle anthropology. For in those studies, he had described Kabylia as 'a society without writing', characterised by 'the absence of politically constituted institutions endowed with the de facto monopoly of legitimate violence' (1980, p. 186 [p. 109]). Further, the politicisation of the Kabyles, the role of colonial or Koranic schools, the history of labour migration between Kabylia and France, to say nothing of the 'thousand-year-old' 'incessant transactions' between the Kabyle and Islamic traditions, had received no mention whatsoever in *Esquisse ... , Outline ... ,* or *The Logic of Practice.*

For Reed-Danahay, such a troubling contradiction can be explained in terms of the role Bourdieu wished Kabylia to play within his wider *oeuvre*. Basing her remarks on a comparative study of *Reproduction* and *Outline ... ,* she argues that Kabylia functioned primarily as a foil to the history of social and cultural domination which Bourdieu had charted in his work on French culture and education. As France's 'exotic other', she maintains, Kabylia represented a realm of relative freedom, of 'possibilities for social manipulation by individuals', which contrasted with the rigid structures of symbolic domination Bourdieu had analysed in his work on French education. According to Reed-Danahay, Kabylia represented a 'utopian vision', 'an ideal model of the traditional with which to rethink our own modern forms of symbolic violence and class culture' (1995, p. 80). She argues that in constructing this 'ideal model', Bourdieu necessarily gave in to 'a form of orientalism', which 'resulted in studies of social and familial practices among the Kabyles to the neglect of national institutions and local-national interactions in that setting' (p. 75). Similar criticisms of Bourdieu's anthropological studies have been voiced by Anthony Free, who argues that Bourdieu's treatment of Kabylia was 'doubly rooted' in colonialism through the 'two absences of the state-political and religious histories of the colonial and pre-colonial history of Kabylia' (1996, p. 409). Camille Lacoste-Dujardin, an anthropologist with many years experience of fieldwork in Kabylia, argues that Bourdieu's approach typifies an 'ahistorical anthropology' with its origins in the French colonial experience (1997, pp. 273–9).

The omissions of historical detail identified by Reed-Danahay, Free and Lacoste-Dujardin are indeed troubling. However, Bourdieu's awareness of the complexities of Kabylia's history, as manifested in his *Libération* article, suggests that those omissions cannot simply be attributed to an unconscious ethnocentrism on his part. Indeed, contrary to what Reed-Danahay has claimed, Bourdieu's understanding of the relationship between Kabylia and the West is by no means reducible to a classically orientalist opposition between idealised 'primitive' utopia and degraded Western modernity. This chapter will seek to clarify that relationship, assessing in greater detail Bourdieu's own accounts of the differences and similarities between Kabylia and the West, of the place of his Kabyle work within his broader intellectual project, and hence of the relationship between the anthropological and sociological aspects of his work.[2] Once the theoretical bases of the relationship between sociology and anthropology in Bourdieu's work have been clarified, it will be possible to assess how this relationship is worked out in practice by examining his 1998 study, *La Domination masculine* (Male Domination). This study provides one of the most condensed examples of the cross-fertilisation of ideas and concepts between Bourdieu's anthropology and sociology, for in it he juxtaposes analyses of gender relations in Kabylia and in the West to suggest certain 'invariant' structures of male domination which cut across any apparent differences between the two societies.

Rehabilitation or Romanticism?

The history of colonial and post-independence Algeria, the 'Kabyle myth', the fact that the terms 'Kabyle' and 'Kabylia' were first coined by French colonial ethnographers to denote a cultural and geographical area whose boundaries they themselves had defined all mean that the very act of writing about Kabylia as a distinct cultural entity is laden with political implications. Thus, even someone sympathetic to the Kabyle cause and critical of the cultural and political hegemony of the FLN, such as the Algerian playwright and novelist Kateb Yacine, can question the use of terms like 'Kabyle' and 'Kabylia', which 'all originate from the enemy powers who have invaded us', which are 'pejorative terms, directed against us', whose use may prove 'suicidal' for post-independence Algeria (in Ouerdane 1990, p. 13). It is surely against this highly charged background that Bourdieu's own reservations concerning the value of anthropological studies of Kabylia can be understood.

In his Preface to *The Logic of Practice*, Bourdieu stated that when he had first started his researches in Algeria he had been suspicious of anthropology's conventional focus on 'traditional' Kabyle rite and ritual, which seemed not only inappropriate, given the circumstances of the war, but also complicit with certain racist or ethnocentric representations of the 'primitive' nature of Algeria's indigenous populations. If, from 1958 onwards, he had pursued the study of 'ritual traditions', it was in order to attempt 'to retrieve' ritual 'from the false solicitude of primitivism and to challenge the racist contempt which, through the self-contempt it induces in its victims, helps to deny them knowledge and recognition of their own tradition' (1980, p. 10 [p. 3]). Bourdieu argued that he had been encouraged in his efforts to rehabilitate Kabyle tradition by 'the interest my informants always manifested in my research whenever it became theirs too, in other words, a striving to recover a meaning that was both their own and alien to them' (p. 11 [p. 3]). Yet, such 'interest' notwithstanding, 'an awareness of the "gratuitous" nature of purely ethnographic enquiry' had continued to haunt Bourdieu, prompting him to publish instead works such as *Travail et travailleurs* ... and *Le Déracinement*, 'two books analysing the social structure of the colonised society and its transformations' (p. 11 [p. 3]). Indeed, although all Bourdieu's fieldwork was conducted over the period 1958–64, with the exception of one article which appeared in an English-language anthology in 1965, he was not to publish his *anthropological* studies of Kabylia until the decade 1970–80, ten years or more after the end of the Algerian War.[3]

Bourdieu's efforts to rehabilitate Kabyle tradition were, of course, entirely laudable and had retained an urgency in the post-colonial conjuncture given the ruling FLN's hostility to non-Arabic languages and cultures. Moreover, this concern to rehabilitate 'traditional' Kabyle society explains the absence from his book-length studies of Kabylia of any mention of the Kabyles' level of politicisation and role in the Algerian War of Independence, of their history of labour migration to France or of their troubled relationship with the ruling FLN in post-independence Algeria. It does not, however, explain Bourdieu's decision not to analyse what, in his article in *Libération*, he described as the 'thousand-year-old' 'incessant transactions' between Kabyle tradition and the instances of Islamic religious and political power.

Further, Bourdieu's use of the present tense throughout *Esquisse ... , Outline ...* and *The Logic of Practice* tended to conceal the fact that the rituals and practices he was describing were no longer current in present-day Kabylia. On occasion, Bourdieu would remind his readers that, as he put it in a footnote to *The Logic of Practice*; 'The narrative present is used here systematically to describe practices that were present in the informants' memories at a given moment but fell into disuse more or less completely, a longer or shorter time ago' (1980, p. 335n.2 [p. 311n.2, trans. modified]). On other occasions, he would invoke his authority as an anthropologist faithfully reporting what he himself had seen and experienced, recounting, for example, the scandal caused by a 'well-known mason', who 'around 1955' demanded cash payment rather than the traditional meal in return for his labour (Bourdieu 1977a, p. 173). At such moments, Bourdieu's account seemed not to be a reconstruction of Kabyle tradition but an analysis of Kabylia as he had experienced it, on the verge of its transition from a pre-capitalist to a capitalist economy. Yet if Bourdieu was attempting to analyse the changing Kabyle society of the 1950s, why did he chose to make no mention of the Algerian War, which had started in 1954 and which was to have such a profound effect on Kabylia?[4]

Bourdieu's manifest wish to rehabilitate Kabyle tradition, to defend it against racist or ethnocentric misrepresentation, cannot, therefore, entirely absolve him of the accusation that he has presented an 'ahistorical' account of Kabylia, as Lacoste-Dujardin and others have charged. Indeed, as Bourdieu himself has pointed out, 'promoting, dignifying or ennobling a particular cultural tradition, in accordance with our desire for "rehabilitation", without however consecrating, canonising, and thus eternalising that tradition' is no easy task; the laudable desire to rehabilitate always carries with it the risk of romanticising or idealising a tradition and hence lapsing into the very ethnocentrism one is seeking to avoid (Bourdieu 1985a, p. 82). If Bourdieu is surely correct in stating there is no easy solution to this dilemma, it could nonetheless be argued that a necessary first step would be to question and possibly abandon anthropology's traditional emphasis on the 'primitive' or the 'traditional', the 'elementary forms' or founding structures of social life in 'exotic', 'other' societies. Indeed, the post-colonial era has seen a whole series of anthropologists attempting to achieve just this by placing the societies they study within the context of broader national and international social, cultural and economic movements.[5]

Bourdieu himself has declared that 'all my work, for more than twenty years, has aimed at abolishing the separation of sociology from anthropology' (1987, p. 92 [p. 74]). Yet his own account of the difference between his earlier studies of Algeria such as *Travail et travailleurs* ... and *Le Déracinement* and his later anthropological studies of 'traditional' Kabyle rite and ritual seemed to depend on retaining a very conventional understanding of the role of anthropology. He seemed to accept, as a matter of course, a definition of ethnographic or anthropological enquiry as consisting of tracking down 'traditional', perhaps even archaic social structures and rituals. This definition of anthropology was then contrasted with a more sociological approach, apparently concerned with the dynamics of social and political change. There would seem, therefore, to be a contradiction between, on the one hand, Bourdieu's claim to have worked through the opposition between anthropology and sociology and, on the other, his continued reliance on a distinction between sociology, as the study of contemporary social phenomena, and anthropology, as the reconstruction of 'traditional' rite and ritual.

Anthropology as 'Socio-Analysis'

In seeking to justify his claim to have 'abolished' the distinction between anthropology and sociology, Bourdieu has pointed to his readiness to apply certain findings, concepts and ideas elaborated in the Kabyle context to his work on French society, culture and education. More specifically, he has emphasised that the 'practical logic' he found to be governing the oppositions between high and low, dry and damp, light and dark, masculine and feminine, which structured the Kabyle world, should not be seen as peculiar to Kabylia. Such oppositions, and the practical logic which governed them, he maintained, were not to be attributed to what he termed 'the essential otherness of a "mentality"' (Bourdieu 1980, p. 40 [p. 20]). On the contrary, he argued that a practical logic, in every way analogous to that which governed the symbolic and spatial oppositions of the Kabyle house, could be found at work behind numerous socio-cultural phenomena in the West.

By way of example, he cited his 1975 article, 'The Categories of Professorial Understanding' ('Les Catégories de l'entendement professoral'), in which he had found a practical logic at work behind the oppositions structuring teachers' and lecturers' comments on their students' work. Far from representing objective criteria of judgement, he argued, such comments – 'brilliant' or 'clumsy', 'subtle' or 'servile', 'ingenious' or 'insipid' – were rooted in class distinctions. Like the gendered oppositions structuring Kabylia, such distinctions were internalised in the habitus of teachers and lecturers to form of a set of 'practical taxonomies'. Inculcated and reinforced through long exposure to the institutions of French higher education, their objective function was to reproduce the unequal distribution of 'cultural capital' and hence naturalise class divisions, just as the function of the practical taxonomies in Kabylia was to naturalise and reproduce the unequal division of labour and prestige between the sexes.

In addition to 'The Categories of Professorial Understanding', Bourdieu cited as further examples his own work in *Distinction* on the oppositions between 'modest' and 'vulgar', 'brilliant' and 'laboured', 'refined' and 'tasteless', which structured 'legitimate' taste, and the respondents to a French opinion poll in 1975 who intuitively associated the Communist Georges Marchais with 'the pine tree, the colour black and the crow', and the right-wing politician Valéry Giscard d'Estaing with 'the oak, the colour white or the lily of the valley' (Bourdieu 1980, pp. 38–9 [p. 20]).[6] In each case, Bourdieu argued, a set of 'practical taxonomies' was at work, a structure of homology and opposition, which was incorporated into the habitus in such a way as to appear to constitute entirely natural or objective systems of classification. These practical taxonomies were, he maintained, rooted in social hierarchies and distinctions, primarily of gender in the case of Kabylia and of class in the case of the West.

However, in an interview published in *In Other Words,* Bourdieu noted wryly that where his analyses of the social and symbolic structures determining gender relations in Kabylia had met with 'approval, or indeed admiration', his analysis, 'of the "categories of professorial understanding" ... , seem like gross transgressions showing a lack of respect for the proprieties' (Bourdieu 1987, p. 36 [p. 25]). These differing responses were, he suggested, indicative of the 'unhealthy' disciplinary distinction between anthropology and sociology. More specifically, Bourdieu argued that to maintain anthropology as a separate academic discipline was necessarily to blunt its potentially critical force. For anthropology, as traditionally understood, was the study of societies which were both geographically and culturally very distant from our own. According to Bourdieu, this very tangible distance, which separated anthropologists and their readers from the societies they typically studied, predisposed them to accept unflinchingly an analysis of the objective determinants of social practice when it applied to some 'primitive' or 'exotic' society. They remained too detached or neutral with regard to anthropological discoveries to acknowledge that, for example, Bourdieu's analysis of gender inequalities in Kabylia might have some relevance to their own societies. As long as the distinction between the two disciplines was kept in place, he argued, this would remain so. For this distinction institutionalised the distance that separates anthropologists from the objects of their studies: 'The distance which the anthropologist puts between himself and his object is institutionalised in the division between anthropology and sociology' (Bourdieu 1980, p. 34 [p. 17]). It allowed the findings of anthropology to be contemplated as mere exotica, attributed precisely to 'the essential otherness of a "mentality"'. As Bourdieu put it:

> The distinction between anthropology and sociology prevents the anthropologist from submitting his own experience to the analysis that he applies to his object. This would oblige him to discover that what he describes as mythical thought is, quite frequently, nothing other than the practical logic we apply in three out of four of our actions: even, for instance, in those of our judgements which are considered to be the extreme accomplishment of cultivated culture, the judgements of taste, entirely founded on (historically constituted) couples of adjectives. (Bourdieu 1987, p. 82 [p. 66])

If the readers and critics of Bourdieu's anthropological work had remained too distant, too objective or detached to grasp its critical force, readers and critics of his sociological work had committed the opposite mistake. For they remained too subjective, too close to the phenomena under discussion to attain the necessary measure of objectivity and critical distance. Indeed, Bourdieu argued, the proximity of these same individuals to the social phenomena which traditionally formed the subject matter of sociology, their pre-reflexive or immediate adherence to the '*doxa*', the apparent self-evidence of their own social world, predisposed them to reject or resist analyses of the objective determinants of their own social practices. Hence, according to Bourdieu, sociological discourse provoked in its readers a response analogous to that elicited by psychoanalysis: 'Sociological discourse arouses *resistances* that are quite analogous in their logic and their manifestations to those encountered by psychoanalytical discourse' (Bourdieu 1980a, p. 41 [p. 23]).

Bourdieu's project was, therefore, twofold. On the one hand, he sought a way of suspending his readers' pre-reflexive investment in everything they took for granted about their own social universe. On the other, he had to work to preserve the critical force of his anthropological work and prevent it being reduced to the study of exotic curios. This twofold project necessarily involved a challenge to the conventional distinction between anthropology and sociology, hence Bourdieu's declaration in 1985 that, 'all my work, for more than twenty years, has aimed at abolishing the separation of sociology from anthropology'. In his sociological work, this implied a constant effort on his part to turn the anthropological gaze on to his own society; an act of de-familiarisation or critical estrangement which aimed to suspend both his and his readers' uncritical investment in the apparent self-evidence of the social conventions which governed their native social universe. This de-familiarising strategy, he claimed, has profoundly political implications: 'The fact of asking of our own societies traditional anthropological questions, and of destroying the traditional frontier between anthropology and sociology, was already a political act' (Bourdieu 1987, p. 36 [p. 24]).

In his anthropological work, Bourdieu needed to achieve the opposite. Namely, he had to render the apparently irremediably 'other' more familiar by working through the distortions inherent in the distance that typically separates anthropologists from the objects of their studies. Not only did this distance threaten to undermine anthropology's critical force, reducing its discoveries to the status of mere exotica, but, according to Bourdieu, it also encouraged the 'objectivism' typified by the work of Claude Lévi-Strauss. This distance could not simply be wished away in an illusory attempt to adopt the native's point of view since this would involve adopting a 'naively phenomenological' mode of social analysis, falling into the diametrically opposed, but entirely symmetrical trap of 'subjectivism'.

The anthropological encounter, as defined by Bourdieu, thus involved the dialectical interplay of these two poles of objectivism and subjectivism, of structuralism and phenomenology, of the strange and the familiar. A sense of the essential otherness of the society under study had to be maintained, even as any similarities between forms of social organisation in Kabylia and those in the reader's or anthropologist's native country were fully acknowledged. In *The Logic of Practice*, Bourdieu described this process as follows:

> So scientific work provides, in this case, a strange experience, bringing the stranger closer without taking away any of his strangeness, because it authorises the closest familiarity with the strangest aspects of the stranger while at the same time imposing a distance – the precondition for a real appropriation – from the strangest aspects of what is most personal. (1980, p. 246 [pp. 145–6])

It was this interplay between the strange and the familiar which endowed anthropology with its critical force. Not only did it prevent the twin errors of objectivism and subjectivism; in staging an encounter with another, distant world, it also forced anthropologists and their readers to stand back from all that they took for granted in their own world, to grasp the objective determinants of their own practices by achieving critical distance vis-à-vis 'the strangest aspects of what is most personal'. Anthropology thus became what Bourdieu termed, by analogy with psychoanalysis, 'a particularly powerful form of socio-analysis':

> Anthropology then ceases to be a kind of pure art that is totally freed, by the distancing power of exoticism, from all the suspicions of vulgarity attached to politics, and becomes instead a particularly powerful form of socio-analysis. By pushing as far as possible the objectification of subjectivity and the subjectification of objectivity, it forces one, for example, to discover, in the hyperbolic realisation of all male fantasies that is offered by the Kabyle world, the truth of the collective unconscious that haunts the minds of anthropologists and their readers, male ones at least. (p. 246 [p. 146])

This quotation invites interpretation on several levels. At its most straightforward, it is a description of the dialectic of estrangement ('the objectification of subjectivity'), and empathy ('the subjectification of objectivity'), which Bourdieu placed at the heart of the anthropological encounter. On another level, it describes Bourdieu's effort to work through the opposition between objectivism and subjectivism, an effort which animates his entire *oeuvre*. Yet these poles of objectivism and subjectivism, the strange and the familiar, also have far more concrete referents. This passage prefaced two chapters on kinship in *The Logic of Practice*, the first of which dealt with Bourdieu's native Béarn, the second with Kabylia. Both drew on fieldwork conducted over the same period in the late 1950s and early 1960s. Thus, at the same time that Bourdieu had conducted his fieldwork in Kabylia, he had also been working on the extended article on marriage and kinship in the Béarnais villages in which he grew up, 'Célibat et condition paysanne' (1962).

Bourdieu argued that it was the experience of working simultaneously on a world that was entirely familiar to him and one that was entirely foreign that first led him to question the adequacy of Lévi-Strauss's 'objectivist' analyses of kinship, and, from there, to reflect on the social determinants of that objectivism, its inherent distortions, and its relationship with the distinction between anthropology and sociology. Analysing matrimonial and kinship strategies in his native Béarn, Bourdieu was able to see the limitations of an objectivist or structuralist analysis of such phenomena. His personal experience of the Béarn meant he saw 'faces behind the statistics, personal experiences, interweaved with shared memories, behind the biographies, landscapes through the cartographic symbols' (Bourdieu 1972, p. 156).

If this might seem to imply a more phenomenological mode of analysis, an appeal to subjective experience, Bourdieu's simultaneous work in Kabylia alerted him to what might be gained by turning the objective, de-familiarising gaze of the anthropologist on to his own society. His two research projects on Kabylia and the Béarn thus formed, 'the crossover point, the interface, between anthropology and sociology' (Bourdieu 1987, pp. 75–6 [p. 59]). Hence, throughout his career, in both his work on his native social universe, be it Béarn or the worlds of French education and 'legitimate' culture, and his work on Kabylia, Bourdieu has attempted to retain the advantages of a typically anthropological distance, even as he reflects on its social conditions of possibility and inherent distortions. In this way, he aims to work through and dialectically transcend the oppositions between objectivism and subjectivism, anthropology and sociology.

Central to this project is the constant alternation between the analysis of a world that is familiar and the study of one that is foreign, between the conventional topics of sociological and anthropological research, respectively, between France and Kabylia. This has led to what Bourdieu termed, 'this sort of alternating experience of the social world, namely the *familiarisation* with a foreign world and the *deracination* of a familiar world which are constitutive of any scientific procedure in the human sciences' (1972, p. 156).

Bourdieu's claim to scientificity and his attempt to think through the opposition between objectivism and subjectivism are, therefore, inextricably bound up with his eagerness to abandon the distinction between anthropology and sociology. For Bourdieu, the anthropological and the sociological aspects of his work stand in a relationship that is both symbiotic and dialectical; the gains of one discipline supplement and transcend the failings of the other, and vice versa. In *Homo Academicus*, he described this relationship as follows:

> The sociologist who chooses to study his own world in its nearest and most familiar aspects should not, as the anthropologist would, domesticate the exotic, but, if I may venture the expression, exoticise the domestic, through a break with his initial relation of intimacy with modes of life and thought which remain opaque to him because they are too familiar. In fact the movement towards the originary, and the ordinary, world should be the culmination of a movement towards alien and extraordinary worlds. But it hardly ever is: in Durkheim, as in Lévi-Strauss, there is no prospect of subjecting to analysis the 'forms of classification' employed by the scholar, and seeing in the social

structures of the academic world ... the sources of the categories of professorial understanding. (1984, pp. 289–91 [pp. xi–xii])

Thus Bourdieu sought to place himself in a distinguished lineage, as a worthy successor to both Durkheim and Lévi-Strauss. Indeed, he seemed here to be pursuing the Durkheimian project of providing a sociological grounding for Kant's a priori categories. If the Kantian allusion had been clear in the title of his 1975 article, 'The Categories of Professorial Understanding', the extended version of that article which formed the first section of Bourdieu's study of the *grandes écoles*, *The State Nobility* (*La Noblesse d'état* 1989), acknowledged the continuing debt to Durkheim. In an obvious reference to Durkheim and Mauss's seminal essay, 'Primitive Forms of Classification' (1901–2), Bourdieu called the section 'Academic Forms of Classification'. More than a mere successor to Durkheim or Lévi-Strauss, however, Bourdieu clearly saw himself as having transcended the limitations of their respective approaches by means of a reflexive return on to the social determinants of the very categories and systems of classification both thinkers applied to their anthropological data.

Bourdieu's understanding of the de-familiarising force of anthropology seemed equally indebted to the phenomenological tradition. For both Husserl and Merleau-Ponty, anthropology offered 'an imaginary variation', a means of breaking with the 'doxic' realm of unquestioned assumptions about one's native social universe to uncover certain 'invariant' forms of human activity. The 'invariant' was, for Husserl, 'a *general essence* ... , the *eidos*, the idea in the Platonic sense', the '*necessary general form*' of a phenomenon which could be grasped through 'the practice of voluntary variation'. If, say, in the course of perceiving a table, I wanted to grasp the essence of perception, I would 'imaginatively vary' the object perceived, imagining I was perceiving not a table but a chair, a house, a dog, and so on. By reflecting on these 'variants', I could then uncover the essence of perception which lay behind the contingent differences between the particular objects perceived: 'in such free variations of an original image, e.g. of a thing, an invariant is necessarily retained' (Husserl 1948, pp. 340–1). As Merleau-Ponty has pointed out, anthropology offered a means for Husserl to achieve this process of imaginary variation, since 'so-called primitive cultures ... offer us variations of this world without which we would remain enmeshed in our preoccupations and would not even see the meaning of our own lives' (1960, pp. 137–8).

Significantly, in explaining the pretensions to general validity of his own social theories, as they range across different fields of enquiry and geographical areas, Bourdieu has invoked this Husserlian practice of imaginary variation: 'Husserl taught that you must immerse yourself in the particular to find in it the invariant' (1992, p. 57 [p. 77]). Bourdieu has sought to historicise Husserl's understanding of imaginary variation, so that it aims not at uncovering the unchanging essences of certain phenomena but rather what he terms certain 'transhistorical invariants' which mean that male domination, say, has taken similar forms at different historical moments and in different cultural contexts. Thus, in *La Domination masculine*, analysis of gender relations in Kabylia would serve as an 'imaginary variation', a means of 'suspending' Bourdieu's and his readers' pre-reflexive adherence to the '*doxa*', to the apparent self-evidence of gender inequalities, in order to uncover 'the invariants which, above and beyond all the changes to the condition of women, can be observed in the relations of domination between the sexes' (1998a, p. 10).

Kabylia as Microcosm and Archetype

Rather than romanticising Kabylia, then, presenting it as 'an ideal model of the traditional', whose more authentic forms of social organisation could be contrasted with a class-divided West, as Reed-Danahay argues, Bourdieu stressed the essential similarities between forms of 'symbolic domination' and 'symbolic violence' in Kabylia and those at work in the West. '*Symbolic* domination' and '*symbolic* violence' were so called because they referred to the process whereby the existing social order was naturalised and reproduced without any recourse to *physical* violence or domination. As we have seen, Bourdieu drew a series of analogies between 'symbolic violence' in Kabylia and the West, between the incorporated 'practical taxonomies' which naturalised and reproduced gender inequalities in Kabylia and those at work in the worlds of French education, politics and culture. Moreover, he pointed to such analogies as evidence of his determination to think through the opposition between anthropology and sociology, and hence to avoid exoticised representations of Kabylia as the West's 'primitive' Other. The most extended and significant of these analogies was the one Bourdieu drew between Kabylia's 'good faith economy', a 'pre-capitalist' gift economy resting on a communal sense of honour, debt and 'disinterested' reciprocity, and the realm of 'legitimate' culture in capitalist societies.

Gift exchange in Kabylia, according to Bourdieu, only *appeared* disinterested, in fact this apparent disinterest concealed the workings of 'symbolic domination' and 'symbolic violence', the struggles to accumulate 'symbolic capital', and hence impose and reproduce arbitrary social hierarchies. In this, gift exchange was analogous to the realm of 'legitimate culture' in capitalist societies, whose apparent disinterest, Bourdieu would argue throughout *Distinction*, masked a series of class-based 'interests' in the accumulation of 'cultural capital'. The 'interests' at stake in both Kabyle gift exchange and the consumption of legitimate culture in capitalist societies had real social effects, they obeyed 'an economic logic' which was, however, not directly reducible to the workings of the cash economy. Bourdieu thus proposed to elaborate 'a general theory of the economy of practices', which would explain the 'economic logic' behind apparently disinterested practices such as gift exchange or cultural consumption. Nakedly economic or monetary exchanges would themselves be merely a subset of this 'general theory of the economy of practices':

> The theory of strictly economic exchanges is a particular case of *a general theory of the economy of practices*. Even when they give every appearance of disinterestedness because they escape the logic of 'economic' interest (in the narrow sense) and are oriented towards non-material stakes that are not easily quantified, *as in 'pre-capitalist' societies or in the cultural sphere of capitalist societies*, practices never cease to comply with an economic logic. (1980, p. 209 [p. 122], my emphasis)

Kabylia was the keystone of this 'general theory of the economy of practices'. As a 'pre-capitalist' society, Kabylia provided Bourdieu with empirical 'proof' of the 'interest' that lay in accumulating forms of capital other than economic capital, an 'interest' which had real social effects without being reducible to the laws of monetary exchange. According to Bourdieu, Kabylia was a society marked by the absence of differentiated, semi-autonomous cultural, economic, juridical and educational fields underwritten by the State, which ensured the convertibility of the different forms of capital invested in those fields. This meant that *only symbolic capital* could be accumulated through the various strategies adopted in Kabylia's 'good faith economy'. The absence of differentiated fields in Kabylia was a key point of difference between it and capitalist societies. For Bourdieu argued that societies with differentiated, semi-autonomous fields were more open to historical change, to 'real events', which emerged at the intersection between those different fields (1987, p. 91 [p. 73]). However, this difference was also a point of similarity.

In 'pre-capitalist' Kabylia, Bourdieu maintained, economic interest could never reveal itself as such but was always concealed behind the communal sense of honour, debt and reciprocity. In capitalist societies, with the emergence of a field of naked economic interests and exchanges, this 'denial of the economy and of economic interest' had found 'its favoured refuge in the domain of art and "culture"', an 'island of the sacred, ostentatiously opposed to the profane, everyday world of production, a sanctuary for gratuitous, disinterested activity' (1980, p. 231 [pp. 134–5]). Hence Bourdieu could argue that the workings of Kabylia's 'good faith economy' were analogous to the workings of the 'disinterested' realm of art and culture under capitalism. Furthermore, Bourdieu argued, 'the development of the capacity for subversion and critique' prompted by this shift from a pre-capitalist 'good faith economy' to 'the most brutal forms of "economic" exploitation' under capitalism, had led to 'a return to modes of accumulation based on the conversion of economic capital into symbolic capital' (1980, p. 230 [p. 133]). Symbolic forms of domination, from the apparently disinterested pleasures of legitimate culture to philanthropy and public relations were increasingly deployed under capitalism as a means to mask the more brutal forms of economic exploitation (pp. 230–1 [pp. 133–4]).

As a pre-capitalist economy governed by symbolic rather than economic exchange, 'the site *par excellence* of symbolic violence', Kabylia could thus offer important insights into the functioning of symbolic domination, whose 'return' as a fundamental organising principle of capitalist societies could then be much better understood. In *The Elementary Forms of the Religious Life* (1912), Durkheim had justified his interest in Australian Aboriginal religions by arguing that social and cultural phenomena which continued to hold force in the West were easier to discern in such 'primitive civilisations' since they constituted 'privileged cases because they are simple cases', offering 'a means of discerning the still present causes upon which the most essential forms of religious thought and practice depend' (1912, pp. 6–8). Kabylia performed an analogous role in Bourdieu's work with regard to symbolic domination and violence; it was a pre-capitalist society in which 'domination can only be exerted in its *elementary form*' (Bourdieu 1980, p. 217 [p. 126]).

Central to Bourdieu's understanding of symbolic domination and violence was, of course, his account of the way in which arbitrary social hierarchies became incorporated into the structures of the habitus as a set of 'practical taxonomies', which never needed to be explicitly articulated in order to be enacted in social practice. At the heart of the habitus was, then, a pre-discursive, pre-predicative relationship to the world, 'that relationship to the world that Husserl described by the term "the natural attitude" or the "doxic experience" , although he omitted to mention its social conditions of possibility' (1998a, p. 15). At the basis of all experience, prior to any judgement or predication, Husserl had identified a pre-reflexive, 'practical' relation to the world he termed '*Urdoxa*' or originary 'doxa' (1948, p. 59; p. 65; p. 387). Reinterpreting this notion of '*Urdoxa*' in sociological terms, Bourdieu suggested that Kabyle society represented an empirical example of 'the state of originary doxa', a state in which the social order 'goes without saying because it comes without saying', in which social norms and conventions, neither written nor codified, were naturalised to such a degree as to pass entirely unquestioned and unchallenged (1977a, p. 167).

Thus Kabylia, as an example of 'the state of originary doxa', represented the archetype of a society in which socially determined criteria of judgement, systems of classification, expectations and dispositions were internalised in agents *without ever needing to be explicitly articulated or objectively codified*. However, Bourdieu has insisted that even in capitalist societies such as France, whose schools and universities clearly represent institutions which inculcate explicitly codified classifications, rules and conventions, such institutions merely work to reinforce dispositions which have been incorporated pre-discursively into each student's habitus from earliest childhood. It is this emphasis on the doxic nature of incorporated social norms and conventions, according to Bourdieu, which distinguishes his work from Marxist theories of ideology. Such theories, he maintains, continue to understand socialisation in terms of 'false consciousness', of a consciously held set of ideas, an ideology susceptible to critique by means of a *prise de conscience*, the moment one becomes aware that the bases of a given ideology are arbitrary and unjust. For Bourdieu, however, the incorporation of social imperatives and norms operates at a much deeper level; incorporated into the dispositional structures of the habitus, such imperatives and norms are placed beyond the sway of a reflexive *prise de conscience*:

> In the notion of 'false consciousness', which certain Marxists invoke to account for the effects of symbolic domination, it is the word 'consciousness' which is superfluous. To speak of 'ideology' is to situate at the level of representations, hence capable of being transformed by that intellectual conversion we call the *'prise de conscience'*, what is in fact situated at the level of beliefs, that is to say at the deepest of our bodily dispositions. (1997, pp. 211–12)

If 'doxic' incorporation operated in capitalist societies, it was in societies like Kabylia, 'without writing' and 'without schools', that this process of pre-discursive, pre-predicative inculcation could be most easily discerned.

Kabylia served Bourdieu, then, as the archetypal instance of the contribution of 'the doxic experience' to the workings of symbolic domination. It represented a microcosm of forms of domination which were also at work in capitalist societies, constituting a kind of restricted arena in which concepts such as 'habitus', 'practice' and 'strategy' could be elaborated and refined before being imported back into analyses of symbolic domination in, say, the field of legitimate culture in France. A series of analogies could thus be drawn between forms of symbolic domination in Kabylia and in France and Bourdieu could point to such analogies as evidence of his determination to dispense with any disciplinary distinction between anthropology and sociology and of the ethnocentrism such a distinction concealed.

But it is important to remember that Kabylia could only play its role as 'the site *par excellence* of symbolic violence', the site of 'originary doxa', as long as it was depicted as a 'pre-capitalist', 'good faith' economy, a society 'without writing' and hence lacking any objectified instances of legitimate power and violence. Not only did this depiction of Kabylia apparently rest on a 'reconstruction' of the 'traditional' forms of Kabyle society, it also seemed to demand the avoidance of any mention either of the thousand-year-old 'incessant transactions' between 'indigenous' Kabyle culture and Islam or of the objectified instances of Islamic and French colonial power, which Bourdieu himself had identified in his 1980 article in *Libération* as the defining characteristics of Kabyle society. Paradoxically, then, the very point where Bourdieu claimed to have overcome the opposition between anthropology and sociology would seem to mark the re-emergence of that opposition. The analogies he had drawn between Kabylia and France relied on the 'reconstruction' of 'traditional' forms of Kabyle social organisation which had 'fallen into disuse more or less completely'. The reconstructive project Bourdieu pursued in his *anthropological* studies of Kabylia thus contrasted strongly with his *sociological* studies of culture and education in France, all of which were concerned with entirely contemporary issues.[7]

Male Domination

If there is one of Bourdieu's works which highlights both the insights and the potential pitfalls attendant on his use of Kabylia in this way, it is surely his 1998 study *La Domination masculine*, itself an extended version of an article of the same name that first appeared in 1990. Bourdieu opened *La Domination masculine* by posing what he termed 'the paradox of doxa', the paradox that, 'with the exception of a few historical accidents', the established social order 'could perpetuate itself so easily', 'its relationships of domination', 'its privileges and injustices' appearing 'so often acceptable and even natural'. Male domination, he suggested, offered 'the example *par excellence* of this paradoxical submission' to an unjust social order, a submission achieved through 'symbolic violence, a gentle violence, imperceptible, invisible to its very victims' (1998a, p. 7). The anthropological analysis of gender relations in a society such as Kabylia, Bourdieu argued, could render this symbolic violence visible by provoking a moment of critical estrangement, by forcing the sociologist and his readers to suspend their pre-reflexive adherence to the apparent self-evidence of gender roles in their native social universe. As he put it:

> only a very particular use of anthropology can enable us ... to objectify scientifically the truly mystical operation which produces the division between the sexes as we know it, or, in other words, to treat the analysis of a society organised through and through according to the androcentric principle (Kabyle tradition) as an objective archaeology of our unconscious, that is to say as the tool for a genuine socio-analysis. (1998a, p. 9)

It was this 'androcentric unconscious', with its 'practical taxonomies', its 'forms of classification' and socially determined 'categories of understanding', its gendered 'vision and division' of the social world, which capitalist societies shared with Kabylia. Analysing the social construction of this unconscious in Kabylia and examining the extent to which its characteristic features had remained in place in Western societies, despite the very real changes to women's formal rights and status, was, Bourdieu argued, to focus on certain 'invariant' features of male domination and hence to 'orient in a different way both research into the condition of women ... and the action destined to transform it' (1998a, p. 124). For Bourdieu, this emphasis on the 'transhistorically invariant' nature of male domination distinguished his approach both from Women's History and from certain feminists who, he maintained, were motivated more by understandable 'militant conviction' than the spirit of genuine scientific enquiry and had hence tended to underestimate the true resilience of the 'androcentric' vision of the social world (p. 10; pp. 121–4).[8]

Crucial to Bourdieu's explanation of what he termed 'the transhistorical constancy of the relation of male domination' was his notion of the 'doxic experience', of the way in which social norms and conventions, a gendered 'vision and division' of the social world were incorporated at the pre-predicative level into the habitus. Kabylia, as the site of 'originary doxa', provided the archetypal instance of male domination as *doxa*, inscribed in the pre-predicative forms of ritual and practice rather than objectified in either institutional or ideological form. Presenting a summary of the analysis of Kabyle gender relations contained in *The Logic of Practice*, Bourdieu held Kabylia up as empirical proof of the fact that 'the male order ... has no need of justification: the androcentric vision imposes itself as neutral and does not need to express itself in discourses seeking to legitimise it' (1998a, p. 15). It was enough for the Kabyles to inhabit a social and symbolic universe overdetermined by a series of gendered homologies and oppositions for them to incorporate these at the pre-reflexive level, into the bodily dispositions of the habitus:

> The action of training, of *Bildung*, in the strong sense of the term, which produces this social construction of the body, only very partially takes the form of an explicit and expressly pedagogic action. For a large part, it is produced automatically, without the intervention of any agent, as the effect of a physical and social order entirely organised according to the principle of androcentric division (which explains the great force of the hold it exerts). Inscribed in things, the male order also inscribes itself in bodies through the tacit injunctions involved in the routines of the division of labour or in collective or private rituals (one thinks, for example, of the prohibitions placed on women's behaviour by their exclusion from male places). (1998a, p. 30)

It was because the 'androcentric vision' was incorporated, and hence naturalised, at this pre-predicative level, Bourdieu argued, that it proved so resilient. Incorporated into the very bodily dispositions of male and female subjects, their hexis and habitus, symbolic domination thus took a form so profound it eluded straightforward rational critique. Those who had internalised the socially constructed identity of the dominated gender experienced that identity 'in the form of *bodily emotions* – shame, humiliation, timidity, anxiety, guilt – or of *passions* and *sentiments* – love, admiration, respect', which betrayed themselves in apparently instinctual reactions such as 'blushing, verbal clumsiness, shaking, anger or impotent rage'. Challenging the social bases of these internalised emotions and apparently instinctual reactions was hence to strike at the very heart of a person's sense of identity and required rather more than a simple act of reflexive consciousness or *prise de conscience*:

> The passions of the dominated habitus (from the point of view of gender, ethnicity, culture or language), itself an embodied social relation, a social law converted into an incorporated law, are not such that they can be suspended by a mere effort of will founded on a liberating *prise de conscience*. If it is completely illusory to believe that symbolic violence can be defeated by the weapons of consciousness and will alone, this is because the effects and conditions of its efficacy are durably inscribed in our bodies at the most intimate level in the form of dispositions. (1998a, p. 45)

Hence, Bourdieu argued, feminists and those influenced by the Marxist tradition, who spoke in terms of 'ideology' and 'false consciousness' and anticipated that 'the emancipation of women' would be 'the automatic effect of a *prise de conscience*', were mistaken;

'for lack of a dispositional theory of practices,' they had overlooked 'the opacity and inertia which result from the inscription of social structures into bodies' (p. 46).

If Kabylia provided an archetypal instance of the way in which social norms and conventions were inscribed into the body, Bourdieu insisted that an entirely analogous process was at work in the United States and Europe. He cited the work of Nancy M. Henley and Frigga Haug on the way young women in the West are trained 'to occupy space, to walk, to adopt suitable body positions' or come to internalise 'feelings linked to the different parts of the body, to the back which must be kept straight, to stomachs which must be kept in, to legs which must be kept together, etc., so many postures charged with a moral meaning' (p. 34). The social construction of women's bodies also functioned, Bourdieu argued, to turn women into the objects of the male gaze, into 'symbolic objects, whose being (*esse*) is a being-perceived (*percipi*)', with the effect that they are placed in 'a permanent state of bodily insecurity or, better, of symbolic dependency' (p. 73).

This process of incorporating the 'dominant taxonomy' as regards gender did not, Bourdieu insisted, apply merely to women as the dominated gender. Men, the dominant, were equally constrained by the roles and identities which, according to the dominant taxonomy, they were supposed to incarnate. Whilst benefiting from their dominant role, men were thus, 'to quote Marx's phrase "dominated by their domination"' (p. 76). The character of Mr Ramsay in Virginia Woolf's novel *To the Lighthouse* (1927) provided Bourdieu with a portrait of a man desperately and somewhat pathetically attempting to live up to the dominant role he was called upon to embody, both literally and figuratively.

Men and women, both in Kabylia and the West, were therefore forced to internalise restrictive gender identities. Bourdieu acknowledged that in the West women's condition had undergone 'profound transformations', as manifest in their increased access to higher education and the labour market, their consequent arrival in 'the public sphere', which had been accelerated by their partial liberation from traditional domestic tasks and reproductive functions, itself facilitated by the availability of contraception and evident in rising divorce rates and the decreasing popularity of marriage (1998a, p. 96). However, he argued that these 'visible changes of *condition* in fact hide continuities in *relative position*' (p. 97). Thus, although women might have gained entrance to higher education and the jobs market, the disciplines they typically studied or the posts they occupied were still in the main far less prestigious than male dominated disciplines and professions. According to Bourdieu the persistence of such inequalities, despite the very real advances that had been made, could be explained by the 'relative autonomy' of 'the economy of symbolic goods', which allowed 'male domination to perpetuate itself' in the symbolic realm, 'over and above any changes to the mode of economic production' (pp. 103–4).

In other words, Bourdieu was arguing that the changes in women's social and economic condition had been mediated through an older, 'traditional' network of assumptions about the role of women, assumptions belonging to the realm of the 'symbolic': 'the changes to the condition of women still obey the logic of the traditional model of the division between the masculine and the feminine' (p. 101). In Kabylia, this 'traditional model' distinguished between the devalued sphere of feminine domestic labour and the valued masculine world of work outside in the fields. This opposition, Bourdieu argued, was reproduced in the distinctions between relatively devalued 'feminine' academic disciplines in the arts and humanities and the prestigious 'masculine' disciplines in the pure sciences, for example, or between those professions which were dominated by women and those that were dominated by men. As he put it: 'The fundamental opposition, which Kabyle society offers in its canonical form, is geared down and, as it were, diffracted in a series of homologous oppositions, which reproduce that fundamental opposition, but in dispersed and often unrecognisable forms (such as the oppositions between the sciences and the arts, or between surgery and dermatology)' (p. 113). If these oppositions had proved so resilient, it was because they operated at the level of the habitus, of those incorporated, pre-predicative structures which were so resistant to rational critique:

> The constancy of the habitus ... is thus one of the most important factors in the relative constancy of the structure of the sexual division of labour. Since the principles [of the dominant vision of the social world] are, in their fundamentals, transmitted from body to body, on the hither side of consciousness and discourse, they escape, in the main, the grasp of conscious control and by the same token are resistant to transformation or correction ... (p. 103)

This account of the way in which 'traditional' gender divisions and inequalities were being reproduced in new forms turned, then, on the notion of the pre-predicative, pre-discursive functioning of the habitus, and much hinged on the ability of Bourdieu's Kabyle data to provide objective empirical proof of the fact that social conventions could indeed be inculcated in agents without ever needing to be expressed in discursive form. Bourdieu's ability to point to his Kabyle data as empirical proof of the 'invariant' nature of male domination was also central to his call for Women's History and certain brands of feminism to turn away from an emphasis on change and progress towards the study of such 'transhistorical invariants', a call which rested on a claim to greater historical accuracy and hence scientificity. However, the empirical or historical validity of Bourdieu's Kabyle data was not itself above question.

In the opening pages of *La Domination masculine*, Bourdieu stated that his analysis of Kabylia represented 'the direct study of a system still in a functioning state' (p. 12). The Kabyles, he maintained, 'have safeguarded, despite conquests and conversions and doubtless in reaction against them, ... structures which, relatively unchanged ... represent a paradigmatic form of the "phallonarcissistic" vision and androcentric cosmology which are common to all Mediterranean societies' (pp. 11–12). However, the notion that Kabylia is still today a society characterised by an unchanging and unchallenged system of gender relations has been strongly contested by Camille Lacoste-Dujardin (1997, pp. 160–71). Drawing on fieldwork she conducted in Kabylia in the 1970s, Lacoste-Dujardin has emphasised the extent to which 'traditional' Kabyle gender relations were radically altered in the post-colonial conjuncture. In a society which had lost a large proportion of its men to the Algerian War or to emigration and in which villages destroyed during the conflict had been rebuilt along more modern, urban lines, women, often financially autonomous thanks to the war pensions of their deceased husbands, had discovered new freedoms and were effectively challenging the old patriarchal system. In an increasingly urbanised, modern Kabylia, it was virtually impossible to find one of those traditional Kabyle houses which Bourdieu had endowed with the central role in the pre-predicative inculcation of gender roles: 'One looks in vain for one of those famous "Kabyle houses" where one would be hard pressed to find the "world reversed" that Pierre Bourdieu wanted to see there' (Lacoste-Dujardin 1997, p. 274).

If, at the opening of *La Domination masculine*, Bourdieu suggested he was presenting an accurate account of contemporary Kabyle society, elsewhere he had stated that his anthropological studies of Kabylia represented 'reconstructions' of its 'traditional' rites and customs, now long since disappeared. As we have already seen, this reconstruction of 'traditional' Kabylia involved the omission of any analysis of the relations between Kabylia and Islam, the French colonial regime, or any wider networks of economic and political power. It is arguable, then, that it was only by screening out any details concerning the institutionalised and codified forms of power in Kabylia that Bourdieu could hold it up as the archetypal instance of 'doxic' or pre-discursive forms of social conditioning at work and, by extension, as representing the 'paradigmatic form of the "phallonarcissistic" vision'. Indeed, even if we take Bourdieu's account of gender in Kabylia on its own terms, it is questionable whether he had really proved that gender roles in that society were inculcated pre-discursively.

For, although he claimed that male dominance in Kabylia 'has no need of justification' (p. 15), and that Kabyle tradition 'is not rich in justificatory discourses', Bourdieu immediately went on to analyse the 'sort of myth of origin' which served 'to legitimise the positions assigned to each sex in the division of sexual labour and, through the intermediary of the sexual division of the labour of production and reproduction, in the social order as a whole and, beyond, in the cosmic order' (p. 24). Further, he insisted that the process whereby young Kabyle males distanced themselves from their mothers to accede to the masculine realm was 'expressly and explicitly accompanied and even organised by the group, which, in all the sexual rites of institution oriented towards virilisation and, more broadly, in all the differentiated and differentiating practices of ordinary existence (sports and virile games, hunting, etc.), encourages the break with the maternal world' (p. 31). By Bourdieu's own account, therefore, there was a series of quite explicit, codified, discursive forms, myths and rites which contributed to the inculcation of gender roles in Kabylia. This raised the possibility that, far from being inculcated into the bodily dispositions of the habitus without ever being expressed in explicit or discursive form, gender roles were, at times, quite explicitly formulated, only subsequently to become internalised, in the form of a kind of sedimentation, into the bodily structures of habitus and hexis.

This argument, of course, would also hold true for questions of gender in capitalist societies. We might agree with Bourdieu that gender identities are often expressed or experienced as a set of implicit, apparently pre-reflexive assumptions and bodily dispositions. However, this does not mean that such identities, their characteristic tics and ways of looking at and being in the world, were internalised at the pre-discursive level, 'on the hither side of words and concepts'. On the contrary, these may be merely the sedimentations of a whole series of discourses, discourses of advertising, of the media, of cinema, of literature of all kinds. In other words, it might be argued that what Bourdieu described so convincingly in *La Domination masculine* as the embodied character of gender identities can be best understood as an example of what Russell Jacoby, echoing Herbert Marcuse, has termed a 'second nature', 'accumulated and sedimented history, ... history that has hardened into nature, ... petrified history' (Jacoby 1977, p. 31).

To distinguish between Marcuse's notion of 'second nature' as history become nature and Bourdieu's account of the pre-discursive incorporation of the *doxa* may seem to be splitting hairs. In fact, however, the distinction is a vital one. For Bourdieu's emphasis on the way social conventions were incorporated into the habitus without ever being explicitly articulated or codified risked transforming the habitus from a structure that had been historically determined into one that was merely culturally arbitrary. Since, in Bourdieu's account, the dispositions incorporated into the habitus need never be mediated through any form of discourse, those dispositions appeared to exist in a realm beyond the grasp of reflexive critique, whilst their incorporation became an immediate, even automatic process. In the case of Marcuse's 'second nature', the process whereby a set of historical determinants have become occluded, frozen or petrified into an apparent 'nature' can itself be reflected upon, criticised and worked through. Much of Bourdieu's argument hinged on the notion that the pre-discursive structures of the habitus were not amenable to this kind of reflexive critique, an argument which left open the question of how those structures might be criticised and transformed.

Having insisted that feminists were wrong to expect that women's emancipation could flow from a 'liberating *prise de conscience*', Bourdieu argued that only 'a radical transformation in the social conditions of the production of dispositions' could challenge the apparent self-evidence of gender roles and identities (p. 48). However, he did not explain how such a transformation might be achieved or who might come to work for such a transformation without having first undergone a *prise de conscience*, without having first become conscious of the arbitrary and unjust nature of male domination. Similarly, Bourdieu acknowledged 'the feminist movement's immense critical labour', which meant that, in the West, male domination no longer went without saying but had to seek to defend and justify itself at every turn (pp. 95–6). Not only did this acknowledgement sit uneasily with Bourdieu's opening claim that male domination was 'the example *par excellence*' of symbolic domination, a form of domination which remained 'invisible even to its victims', it also left unanswered the question of how feminists had achieved this 'critical labour' without themselves having undergone a 'liberating *prise de conscience*'.

In conclusion, if Bourdieu's account of the centrality of certain bodily processes to *the experience* of male domination was both entirely plausible and an important contribution to the analysis of gender issues, his insistence that such domination was internalised by agents without ever needing to be expressed in discursive form seemed less convincing and risked presenting an unchanging and unchangeable vision of gender relations. Ultimately, the plausibility of Bourdieu's analyses of gender hinged on his use of the anthro-pological data he had first presented in his studies of Kabylia. As such, *La Domination masculine* nicely encapsulated the various roles which Kabylia has played in Bourdieu's work to date. In the first instance, the turn to Kabylia was intended to provoke a kind of de-familiarisation or critical estrangement in Bourdieu's readers. This estrangement, however, did not aim to recapture some more authentic, 'primitive' forms of social existence lost to Western modernity. In keeping with Bourdieu's determination to eschew exoticising representations of the 'primitive', and hence to work through the opposition between sociology and anthropology, Kabylia provided, rather, an archetype of forms of symbolic domination and doxic adherence discernible also in the West. Most importantly, as a 'society without writing' and 'without schools', Kabylia was intended to provide empirical proof as to the importance of pre-discursive forms of social conditioning, the very forms of social conditioning whose importance, Bourdieu argued, Marxists and feminists alike had tended to overlook.

However, as critics such as Reed-Danahay, Free and Lacoste-Dujardin have noted, to portray Kabylia as a society 'without writing' and 'without schools' was to risk ignoring details of the region's relationship with wider networks of religious, political and economic power. The ability of Kabylia to provide empirical proof as to the importance of pre-discursive forms of social control and action thus seemed seriously compromised and it appeared that the historical specificity of the region was being subordinated to the role Bourdieu wanted Kabylia to play within his wider social theory, as the site of 'originary doxa', the archetypal instance of symbolic domination and violence, the keystone, in short, of his 'general theory of the economy of practices'.

It was in the light of this general theory of the economy of practices that Bourdieu had returned to an analysis of the links between aesthetics, taste, culture and class in *Distinction*, originally published in France in 1979. As Bourdieu argued in *In Other Words*, *Distinction* and *The Logic of Practice* constituted 'two complementary books' (1987, p. 33 [p. 23]). Didier Eribon (1980, p. 1), meanwhile, has suggested that the two works form 'one and the same work .... *The Logic of Practice* establishes the theoretical foundation of the empirical researches presented in *Distinction*'. Given that the empirical core of *Distinction* was an extended version of the survey into tastes and cultural consumption which had formed the basis of *Photography*, the next chapter will examine to what extent Bourdieu's general theory of practice, and all the problems inherent in its elaboration, inflected and altered his earlier analyses of the relationship between class and culture.

Notes to Chapter 5

1. For a more detailed account of the 'Kabyle myth', see Ageron (1968) and Lorcin (1995).
2. Bourdieu conventionally uses the terms '*ethnologie*' or '*ethnologique*' to describe his Kabyle work. These have been variously translated as 'ethnology/ethnological' or 'anthropology/anthropological'. For the

sake of consistency, I have used 'anthropology' and 'anthropological' in this chapter.

3. The article in question was entitled 'The Sentiment of Honour in Kabyle Society' and appeared first in the 1965 anthology, *Honour and Shame: the values of Mediterranean society* (see Bourdieu 1965c). This essay was later to form one of the 'three studies of Kabyle anthropology' which opened *Esquisse* ....

4. For a moving personal account of the effects of the war on Kabylia, see Feraoun (1962).

5. For one recent example of this approach, see Geertz (1995).

6. Bourdieu was referring to his 1976 article, 'Un Jeu chinois: notes pour une critique sociale du jugement', which he subsequently included in an Appendix to *Distinction*.

7. One obvious exception to this was Bourdieu's study of the nineteenth-century literary and artistic fields, *The Rules of Art* (*Les Règles de l'art* 1992). However, this was identified as a work of historical research referring to a clearly defined historical epoch in a way in which his anthropological studies of Kabylia were not.

8. Bourdieu's critique of feminism and his claim to speak for objective social science, which he first voiced in the article 'La Domination masculine' in 1990, has been subjected to a surely justified critique by a group of French feminists of the materialist school, associated with the journal *Nouvelles questions féministes* (see Armengaud et al. 1995).

You must be _OF_ the society to understand it (sociology) but must maintain a distance in analysis to ensure objectivity and reduce inherent subjectivity that comes w/ familiarity (anthropology)

CHAPTER 6

Old Wine, Distinctive New Bottles

In 1979, Bourdieu published what was to prove his most detailed and surely most influential study of the links between class, culture and social reproduction. *Distinction* was the culmination of over fifteen years' work, having its origins in the survey of tastes and patterns of cultural consumption Bourdieu had conducted in 1963 amongst a sample of 692 inhabitants of Paris, Lille and an unnamed provincial town. He had already used some of the data from this survey in his contribution to the 1965 study, *Photography*. However, much of the more general data had remained unused and in 1967–68 he had conducted a second stage of research, boosting his sample to a total of 1,217 respondents. It was these data, supplemented by numerous surveys of cultural and leisure activities, which formed the empirical core of *Distinction*.

If Bourdieu's concern with the relationship between class and cultural consumption remained constant from his early studies, such as *Photography* (1965) and *The Love of Art* (1966), to the later *Distinction*, the interpretative framework he brought to bear on this issue had undergone considerable refinement and modification. The key concepts of 'habitus', 'strategy' and 'practice' were to be employed in a way which had remained underdeveloped in the earlier works, whilst cultural practices in contemporary France were to be understood in terms of the 'general theory of the economy of practices', which Bourdieu had elaborated in his studies of Kabylia. This was made clear by the very first sentence of *Distinction*: 'There is an economy of cultural goods, but this economy has a specific logic which must be grasped in order to avoid economism' (1979, p. I [p. 1, trans. modified]). Just as the Kabyle gift economy functioned according to a logic which was not directly reducible to the workings of a cash economy, so, Bourdieu argued, the economy of cultural goods in the West functioned according to its own specific, 'practical logic'.

It was not merely Bourdieu's interpretative framework which had changed in the interim between the publication of his early works on class and culture and *Distinction*. The intellectual, political and socio-economic context against which the earlier works had been published had been transformed. Politically, the legacy of May 1968 was manifest in the resignation of Charles de Gaulle in 1969 and the efforts of subsequent presidents, Georges Pompidou and Valéry Giscard d'Estaing, to appropriate some of the demands of May to their own, right-wing, political programmes. Economically, the first oil crisis of 1973–74 had signalled the end of the thirty-year period of French postwar economic growth and reconstruction, and heralded a much less optimistic socio-economic conjuncture. From the late 1970s onwards, France, in common with other major Western economies, was to suffer a period of low growth and high inflation, whilst being burdened with a high rate of structural unemployment. In terms of the French intellectual field, the 1970s saw the publication of some of the most influential 'poststructuralist' or 'postmodernist' texts: to cite but a few examples, Gilles Deleuze and Félix Guattari's *Anti-Oedipus* (1972), and Jean Baudrillard's *For a Critique of the Political Economy of the Sign* (1972) and *The Mirror of Production* (1973), had all appeared in France in the 1970s; Jean-François Lyotard's *The Postmodern Condition* was first published in France in 1979, the same year as *Distinction*.

This chapter will start by outlining how Bourdieu's earlier analyses of the links between class, culture and aesthetics were reinterpreted in terms of his now fully developed 'theory of practice'. Once Bourdieu's general model of the various class-determined aesthetics has been elucidated it will be possible to show how he complicated that initial model by taking full account of the impact of the changing economic, political and social conjuncture of post-1968 France. Close analysis of Bourdieu's account of the impact of this new conjuncture on aesthetic judgements and class distinctions will also enable a much clearer assessment of his relationship to 'postmodernist' or 'poststructuralist' modes of thought.

The Aesthetic Disposition as 'a Practical Sense'

In *The Rules of Art*, Bourdieu was to distinguish between two stages in his understanding of the processes involved in the appreciation of a work of art. His early work, he suggested, remained too 'intellectualist', focusing on art appreciation in terms of the possession or otherwise of the requisite hermeneutic code with which to decipher the work in question. Progress beyond this 'intellectualist' mode of analysis was aided by two key developments. Firstly, his elaboration of a theory of practice in his Kabyle work and, secondly, his reading of Michael Baxandall's study, *Painting and Experience in Fifteenth-Century Italy* (1972). The combination of these two influences, Bourdieu argued, allowed him to understand art appreciation less as an intellectual exercise of decipherment than as a social *practice*; it encouraged him to think in terms less of a codified set of knowledge than of an aesthetic *disposition* and to analyse that disposition in terms of 'the specific logic of practical sense, of which aesthetic sense is a particular case' (1992a, pp. 431–4 [pp. 313–15]).

Baxandall's study demonstrated how little choice individual artists had had over the colour, size, form and subject matter of early Renaissance paintings, such features being prescribed by the prince, merchant or priest who had commissioned them. Art in the early Italian Renaissance, Baxandall argued, was produced and consumed in accordance with a set of communal social, moral and religious values and hence was seen as fulfilling a clear social function. Such a 'functionalist' conception of the artist's role contrasted strongly with modern notions of the artist as an original creative genius and suggested a way of responding to art very different from the attitude of 'disinterested' aesthetic contemplation described by Kant in *The Critique of Judgement*, that founding text of modern aesthetics.

As in his earlier work, Bourdieu took *The Critique of Judgement* as the paradigm of 'legitimate aesthetics'. He was able to draw on Baxandall's study to point to the historically and culturally arbitrary nature of Kant's notion of 'disinterested' aesthetic contemplation. Only with the emergence of the figure of the autonomous artist, working in a field of autonomous artistic production, did works of art themselves demand to be appreciated autonomously, by that 'disinterested' gaze demanded by Kantian aesthetics of the 'true' aesthete (1979, pp. III–IV [pp. 3–4]). Bourdieu's enquiries into the historical genesis of this field of autonomous artistic production were to culminate in the publication of *The Rules of Art*, his study of the French literary and artistic fields at the time of Manet, Flaubert and Baudelaire. In *Distinction*, he was concerned less with the production than with the consumption of cultural artefacts, arguing that the ability to adopt the 'pure' gaze demanded by legitimate aesthetics was socially determined, the expression of a habitus inculcated in very specific historical, social and economic circumstances.

In their fundamentals, Bourdieu's findings on the class-based nature of aesthetic tastes echoed those he had already presented in his earlier works on class and culture. Thus, for example, he argued that his data on the aesthetic preferences and cultural practices of the different social classes revealed that the bourgeoisie was statistically by far the most likely to adopt the attitude of disinterested contemplation demanded by legitimate aesthetics. According to Bourdieu, this disinterested aesthetic response was merely the expression of a class ethos or habitus, itself ultimately determined by the distance or immediacy of material need. To be able, as the dominant class was, to contemplate a work of art on the level of form alone, with no concern for its function or practical utility, reflected an attitude of leisurely, contemplative distance on the world dependent on the distance from immediate material need charac-teristic of bourgeois experience. Bourdieu argued that to contemplate a work of art with the 'pure' or 'disinterested' gaze demanded by legitimate aesthetics was to stage a 'break', 'rupture' or 'epoche' with the ordinary world, which was itself a 'social break': 'The pure gaze implies a break with the ordinary attitude to the world, which, given the conditions in which it is performed, is also a social break' (1979, p. V [p. 4]). The aesthetic disposition formed part of a more 'generalised capacity' on the part of the bourgeoisie to stage a 'break' from or 'bracket off' the realm of ordinary existence and material necessity, a capacity which presupposed 'the distance from the world ... which is the basis of the bourgeois experience of the world' (p. 57 [p. 54]).

The bourgeoisie's distance from the realm of immediate material necessity, argued Bourdieu, determined their whole class ethos, an ethos which in turn functioned as 'the generative formula' determining tastes and practices not merely in the rarefied domain of aesthetics, but also in such mundane areas as clothing, food, interior decor, sport and leisure. He admitted that the ability to appreciate a work of art required a certain cultural competence, an inherited stock of knowledge and reference, a code with which to decipher its meaning (p. III [p. 3]). However, he insisted that this 'inheritance' was less the product of formal or explicit education than the expression of a structure of dispositions incorporated 'on the hither side of discourse': 'Bourgeois culture and the bourgeois relation to culture owe their inimitable character to the fact that ... they are acquired, on the hither side of discourse, by early immersion into a world of cultivated *people, practices* and *objects*' (p. 81 [p. 75]). To be a connoisseur, Bourdieu argued, required a 'practical mastery' of the principles of aesthetic taste and judgement, which could not be transmitted 'solely by precept or prescription' (p. 71 [p. 66]). Just as 'practical mastery' of the social norms and conventions in Kabyle society defined '*le paysan accompli*', the complete or accomplished peasant, so in the West, 'the judgement of taste is the supreme manifestation of the discernment which ... defines the accomplished individual [*l'homme accompli*]' (p. 9 [p. 11]).

The bourgeois aesthetic was, then, analogous to the 'practical sense' of the Kabyle villagers. It was less the product of formal education than a structured set of dispositions, a habitus inculcated at the pre-predicative level, the expression of the experience of being born into and inhabiting a cultured or refined spatial, social and affective universe. Working-class people, Bourdieu maintained, inhabited a radically different social and affective universe; their class ethos and their aesthetic thus contrasted strongly with that of the bourgeoisie. The working-class ethos was determined by the collective experience of material necessity and expressed itself in a realist aesthetic, a preference for function over form, quantity over quality, the 'straightforward' or immediate pleasures over the 'refined' or deferred, whether in the domain of art, sport, food or fashion. This realist aesthetic also had its origins not so much in formal education or its absence as in the incorporated structures of the habitus. Bourgeoisie and working class possessed a set of antagonistic 'practical taxonomies' which reflected the very different social environments in which they had been inculcated:

> If a group's whole lifestyle can be read off from the style it adopts in furnishing or clothing, this is not only because these properties are the objectification of the economic and cultural necessity which determined their selection, but also because the social relations objectified in familiar objects, in their luxury or poverty, their 'distinction' or 'vulgarity', their 'beauty' or 'ugliness', impress themselves through bodily experiences as profoundly unconscious as the quiet caress of beige carpets or the thin clamminess of tattered, garish linoleum, the harsh smell of bleach or the perfumes as imperceptible as a negative scent. (p. 84 [p. 77])

If the working class's relationship to material necessity generated a particular class ethos and through that a particular aesthetic disposition, it also generated a very different conception of which qualities constituted '*l'homme accompli*'. According to Bourdieu, the working-class ethos was based on the collective experience of material necessity; making a virtue of that necessity, their ethical code valued 'honesty' and 'straightforwardness' over the niceties and formalities of bourgeois social convention, it favoured the physical force of the manual worker, weight-lifter or wrestler over the deft grace of the tennis player or fencer. Thus Bourdieu described 'two antagonistic world views, two worlds, two representations of human excellence'. Neither of these antagonistic visions was neutral or absolute; the classifications and forms of behaviour valued by one class would be rejected by the other as either too 'vulgar' or too 'formal', depending on which class perspective was adopted:

> what is for some shameless and slovenly, is for others straightforward, unpretentious; familiarity is for some the most absolute form of recognition, the abdication of all distance, a trusting openness, a relation of equal to equal; for others, who shun familiarity, it is an unseemly liberty. (p. 222 [p. 199])

Caught between these two visions of the world lay the petty bourgeoisie. Bourdieu argued that their class ethos was determined by their need to distinguish themselves from the working class and their aspirations to social betterment. Amongst the more traditional fractions of the petty bourgeoisie, shopkeepers, artisans and the like, this would manifest itself in a rigorous work ethic and a rejection of both the 'frivolity' of bourgeois culture and the 'vulgarity' of the working class. Less conventionally moralistic, Bourdieu argued, were the newer petty-bourgeois fractions, the upwardly mobile *classes moyennes*' who had benefited from the expansion in higher education and were taking up white-collar posts in the growing tertiary sector. Their aspirations manifested themselves in their 'cultural goodwill', that enthusiasm for a culture too recently acquired, a culture which still bore the visible marks of the effort involved in its acquisition and could not, therefore, compete with the casual self-assurance of the 'natural' aesthete. Where the petty bourgeoisie sought to distinguish itself from working-class 'vulgarity', the bourgeoisie sought to distinguish itself from petty-bourgeois 'pretension':

> every petty-bourgeois profession of rigour, every eulogy of the clean, sober and neat, contains a tacit reference to uncleanness, in words or things, to intemperance or improvidence; and the bourgeois claim to ease or discretion, detachment or disinterestedness, need not obey an intentional search for distinction in order to contain an implicit denunciation of the 'pretensions', always marked by excess or insufficiency, of the 'narrow-minded' or 'flashy', 'arrogant' or 'servile', 'ignorant' or 'pedantic' petty bourgeoisie. (p. 274 [pp. 246–7])

According to Bourdieu, then, questions of taste, culture and lifestyle were played out on an antagonistic field of struggle between the classes: 'the definition of art, and through it the art of living, is an object of struggle among the classes' (p. 50 [p. 48]). Individuals and classes were portrayed as being involved in a constant process of judging and being judged, of classifying and being classified, of defending the classifications they valued against those valued by other classes. Class identity was as much about a shared set of tastes and aversions as it was about a shared relationship to the relations of production: 'it is, finally, an immediate adherence, at the deepest level of the habitus, to the tastes and distastes, sympathies and aversions, fantasies and phobias which ... forges the unconscious unity of a class' (p. 83 [p. 77]). Drawing on data concerning the consumption of food, clothing, music, film, art and sports, Bourdieu argued that lifestyle 'choices', from the kind of car driven or holiday taken, to the newspaper read or the interior decor chosen, formed 'the small number of distinctive features which, functioning as a system of differences, of differential deviations, allow the most fundamental social differences to be expressed' (p. 249 [p. 226]).

Bourdieu's emphasis on a system of classification made up of differential terms revealed his continuing debt to structuralism.

However, this was a structuralism reformulated in accordance with the 'theory of practice' he had elaborated in his Kabyle work. In an article published five years before *Distinction*, 'Avenir de classe et causalité du probable' (Class Futures and the Causality of the Probable) (1974), Bourdieu had offered the following schematic representation of what he termed 'the dominant ethical taxonomy':

Table 1. The Dominant Ethical Taxonomy (Bourdieu 1974, p. 26)

(BOURGEOIS):	(PETTY BOURGEOIS):	(THE PEOPLE):
'distinguished'	'pretentious'	'modest'
comfortable, ample (of spirit, gesture, etc.)	narrow, repressed, sham	*gauche*, ponderous, embarrassed, timid, clumsy
generous, noble, rich	small, mean-minded, stingy, parsimonious	'awkward', poor, 'modest'
broad-minded liberal, free	strict, formal, severe	'simple', 'childlike', 'nature'
flexible, natural, at ease, relaxed, assured	rigid, tense, restrained	frank, straight-talking
open, broad, etc.	scrupulous, precise, etc.	solid, etc.

Like the oppositions between inside and outside, light and dark, masculine and feminine structuring the Kabyle world, Bourdieu argued that these differential terms obeyed a 'practical logic'. The meaning and value of each term varied depending on one's class perspective, so that language and the classifications it mobilised became the locus of a struggle between each class's conflicting ethos:

Destined to function in practice, in the service of practical functions, [this taxonomy] obeys a *practical logic*. Thus, the 'people' which the 'bourgeois' or rather the dominant fractions of the dominant class conjure up when they think of the 'people' in opposition to the petty bourgeoisie is not the 'people' they produce when they think of them in opposition to the urban worker. No more is it the 'people' conjured up in the populist imagination (more widespread amongst intellectuals, the dominated fractions of the dominant class) in opposition both to the 'bourgeois' and the 'petty bourgeois', that is to say the good honest 'proletarian', hearty, simple, frank, solid and generous, separated by only a few inversions of the meaning of each sign from the modest and clumsy worker of the conservative imagination. (1974, p. 26n.35)

Judgements and classifications of taste, and the language used to express them, were thus, according to Bourdieu, shot through with conflicting class and ideological values. Language was, to use Mikhail Bakhtin's term, inherently 'double-voiced', the same word taking on a radically different valency depending on the context of its utterance:

Behind their apparent neutrality, words as ordinary as 'practical', 'sober', 'clean', 'functional', 'amusing', 'delicate', 'cosy', 'distinguished' are thus divided against themselves, because the different classes either give them different meanings, or give them the same meaning but attribute opposite values to the things named. (Bourdieu 1979, p. 216 [p. 194])[1]

The significance or value attributed to each of the terms in the dominant ethical taxonomy was thus both arbitrary and differential. However, this did not prevent some of those terms, and the cultural practices they named, from having greater objective legitimacy than others. This signalled a significant difference between Bourdieu's analyses of cultural practices in Kabylia and those in the class-divided West. He argued that where in Kabylia there had only been one communally recognised conception of those values which constituted the '*paysan accompli*', in a class-divided society the values which constituted the legitimate conception of '*l'homme accompli*' would be those of the economically dominant class, the bourgeoisie. Elevated to the level of an aesthetic and ethical norm, the culture and ethos of the bourgeoisie could then constitute itself as so much 'cultural capital', a means of legitimating that class's economic and political domination:

This is the difference between the legitimate culture of class-divided societies, a product of domination predisposed to express and to legitimate domination, and the culture of little-differentiated or undifferentiated societies, in which access to the means of appropriation of the cultural heritage is fairly equally distributed, so that culture is fairly equally mastered by all members of the group and cannot function as cultural capital, i.e., as an instrument of domination, or only so within very narrow limits and with a very high degree of euphemisation. (1979, pp. 252–3 [p. 228])

Thus, Bourdieu argued, unlike the ethos of the working class or the petty bourgeoisie, the bourgeois ethos was legitimated in a philosophical discourse of aesthetics which elevated the typically bourgeois attitude of leisurely, contemplative distance on the world to a prescriptive norm of apparently universal validity. He argued that legitimate aesthetics was 'a well-founded illusion' inasmuch as it reflected, at the level of philosophical discourse, the bourgeois experience of the world. However, in positing a socially acquired aesthetic disposition as a prescriptive norm of universal validity, it elevated a class-specific experience to the level of a 'natural' measure of intellectual and spiritual worth, hence naturalising and legitimising class distinctions. One of Bourdieu's primary aims in *Distinction* was to debunk the claims of legitimate aesthetics to universal validity by demonstrating that lofty assertions of aesthetic 'disinterest' were in fact rooted in profoundly material and social 'interests', that the same principles could be found at work behind a taste for Picasso as a taste for *pâté de foie gras*, to use Elizabeth Wilson's example (Wilson 1988). By effecting this 'barbarous reintegration of aesthetic consumption into the world of ordinary consumption' (1979, p. 110 [p. 100]), Bourdieu sought to challenge what he termed 'class racism', that process whereby the bourgeoisie's socially determined taste for legitimate culture was passed off as a marker of their inherent moral and spiritual superiority and, by extension, of the dominated classes' inherent inferiority.

The Changing Political and Economic Context

By reinterpreting his findings on class and culture in terms of the 'theory of practice' he had elaborated in his studies of Kabylia, Bourdieu was thus able to offer a more dynamic account of the social field, a field seen now as marked by constant struggles between the different social classes over questions of language, taste and judgement. However, *Distinction* did not only focus on the conflicts *between* the different social classes concerning which cultural and aesthetic values should predominate. At the core of Bourdieu's argument was his contention that there was a struggle going on *within* the dominant class itself over such issues. At stake in this struggle was a wholesale redefinition of what it was that should constitute *legitimate* culture.

Several English-speaking critics have criticised Bourdieu's conception of legitimate culture in *Distinction* for drawing too heavily on the French experience and exaggerating the importance of *belle lettriste* culture over more commercial or managerial forms of knowledge and aptitude (Giddens 1986; Jenkins 1992, p. 148; Lamont 1992; Shusterman 1992, pp. 196–7; Erickson 1996). What such criticisms ignore is the extent to which *Distinction* anticipated and sought to trace precisely those shifts in the educational, economic and cultural fields which accompanied the increasing ascendancy of commercial over high cultural values. Bourdieu emphasised that the ascendancy of cultural over economic capital, or vice versa, was itself 'at all times a stake in struggles', and that if in certain conjunctures, 'cultural capital may be, as in present-day France, one of the conditions for access to control of economic capital', these conjunctures could and did change (1979, pp. 131–2 [p. 120]). Indeed, he argued that two opposing visions of legitimate culture, one based primarily around literary or artistic values, the other around commercial values, had existed for many years. Currently, commercial values were challenging the hegemony of artistic or literary values (pp. 102–3 [p. 93]). The opposition between the admirable 'disinterest' of the intellectual and the narrow 'materialism' of the bourgeois businessman was thus giving way 'to the opposition between the gratuitous, unreal, unrealistic culture of the intellectual and the economic or polytechnical culture of "modern managers"'. If *belle lettriste* culture still enjoyed such prestige in France, this was attributable to 'the hysteresis of habitus', the time lag between changing material circumstances and their impact on the systems of classification and judgement contained in each agent's habitus (p. 361 [p. 315]).

Bourdieu's analysis and critique of the increasing hegemony of commercial and managerial culture in *Distinction* needs to be understood in the context of contemporary shifts in French politics, economics and society. Bourdieu had started to trace these shifts in his work on class, culture and education in the 1960s. However, such shifts had grown more pronounced in the post-1968 conjuncture and hence occasioned a more detailed critique on Bourdieu's part, a critique initiated in *Distinction* and pursued further in his work of the 1980s and 1990s. In Chapter 2, we saw that Bourdieu's early work on class and culture was set against a background of rapid economic growth and that it sought to challenge the claims made regarding the effects of that growth on French society by both 'technocrats' and the 'massmediologists' close to the French New Left. In the post-1968 conjuncture, Bourdieu's assertions regarding the complicities between these two apparently opposed groups seemed to have been proved correct.

In his 1967 article, 'Sociology and Philosophy in France since 1945: Death and Resurrection of a Philosophy without Subject', one of Bourdieu's primary targets had been those 'meeting places', such as the Club Jean Moulin, where high-ranking civil servants or technocrats met and exchanged ideas with politicians and sociologists of the French New Left. In this context, one could cite figures such as the sociologist Michel Crozier, the civil servant and leading member of the CFDT, the Christian Socialist trade union, Jacques Delors, or the figurehead of the PSU, the moderniser

Pierre Mendès France. Following the resignation of De Gaulle in 1969, it was these figures and their ideas which were to be taken up by the new President, Georges Pompidou's first prime minister, the right-winger Jacques Chaban-Delmas. Chaban-Delmas's project for 'a new society', itself an attempt to capitalise on the desire for change expressed in May 1968, was greatly influenced by the ideas of Crozier and the Club Jean Moulin and was undertaken in close collaboration with Delors and the *mendésiste* Simon Nora, both civil servants at the Ministry of Finance (Berstein and Rioux 1995, p. 31; p. 51).

Further confirmation of Bourdieu's assertions regarding the complicities between the French Right and certain sections of the New Left seemed to have been provided in the support of Jean-Jacques Servan-Schreiber and Françoise Giroud for the right-winger Valéry Giscard d'Estaing, the successful presidential candidate in 1974. Co-founders of the magazine *L'Express*, which had played a key role in France's postwar modernising effort, Servan-Schreiber and Giroud had been passionate supporters of Mendès France, before rallying to Giscard and being offered ministerial posts in his first government (Becker 1998, p. 98).

Giscard had been elected French president just as the effects of the first oil crisis began to be felt. The liberalism of his politics in terms of social policy, as summed up in the electoral slogan 'change without risk', represented another attempt by a right-wing politician to channel some of the reforming impetus of May 1968. Giscard's economic liberalism, meanwhile, reflected the less optimistic economic conjuncture heralded by the oil crisis. As Emmanuel Godin (1996) argues, the economic crisis of the mid-1970s provoked some of France's governing elite to re-think the French state's *dirigiste* role in economic planning and to take the first tentative steps towards an embrace of more liberal or neo-liberal economics. This interest in liberal ideologies was fuelled by the election of Giscard d'Estaing, who affected a more cosmopolitan, dynamic, modern style of presidency which, ostensibly at least, looked to America for inspiration rather than to the Gaullist tradition of centralised economic planning.

It would be wrong to exaggerate the significance of this political shift; neo-liberalism never achieved the hegemony it enjoyed in Britain and America after the election of Margaret Thatcher and Ronald Reagan, in 1979 and 1980 respectively. Indeed, Suzanne Berger (1987, pp. 85–6), who notes the 'first signs' of 'this new liberal synthesis' in the mid-1970s, emphasises that the 'years of the Giscard presidency are today described by businessmen as a time of maximum interference in their day-to-day decisionmaking by a handful of top civil servants close to the President'. Jack Hayward (1986, p. 34) has talked of 'a change to a new model "entrepreneurial state" in France', at this time, 'enveloped' in Giscard's 'liberal rhetoric'. It would perhaps be better to think in terms of a French political and administrative elite who continued to exert considerable influence over the workings of the French economy, primarily through the management of large partly or wholly state-owned enterprises, whilst increasingly adopting a discourse championing a certain economic liberalism, managerial efficiency and the imperatives of international business.

It was the structure and historical genesis of this mix of technocratic *dirigisme* and liberalism that Bourdieu had traced and satirised in a lengthy article of 1976, co-authored with Luc Boltanski, 'La Production de l'idéologie dominante' (The Production of the Dominant Ideology). Expanding on the analysis first offered in 'Sociology and Philosophy in France since 1945', Bourdieu outlined the emergence of this 'dominant ideology' over the *longue durée*, locating its roots in the interwar years in groups such as *X-Crise* or the personalities collected around Emmanuel Mounier and the journal *Esprit*, and tracing its development amongst postwar administrators associated with the Plan, such as François Bloch-Lainé and Claude Gruson, intellectuals, such as Servan-Schreiber, Crozier and Jean Fourastié, and politicans, such as Giscard, Michel Poniatowski and Lionel Stoléru. Bourdieu emphasised that this new 'dominant ideology' was not the preserve of the Right, citing Delors as a prominent Socialist who had rallied to this 'technocratic "left-wing" utopia' based on the notion of 'democratic planning' (1976, p. 37). However, he did argue that it should be seen as a 'reconverted' or 'progressive conservatism', which had replaced older forms of conservatism and was enabling the bourgeoisie to retain its dominance in a modern multinational economy:

> Because reconverted conservatism chooses what is necessary for the conservation of the established order, that is to say economic (and even 'social') progress, it defines itself against primary conservatism, which performs a final service by allowing reconverted conservatism either to pass unnoticed or to appear progressive. [...] Progressive conservatism is characteristic of a fraction of the dominant class which takes as its *subjective law* what constitutes the objective law of its continued dominance, the need to change in order to conserve. (1976, pp. 42–3)

Giscard's brand of liberalism is often understood as deriving from an 'Orléanist' current in French right-wing politics, a form of *laissez-faire* economics and social liberalism markedly different from the 'Bonapartist' tradition incarnated by De Gaulle, for example (Rémond 1982, pp. 290–312). Writing in the interwar years, Maurice Halbwachs had distinguished between the 'traditional' French bourgeoisie and a new stratum of entrepreneurs, who had first emerged during the reign of Louis Philippe d'Orléans (1830–48): 'less impervious to fresh influences', inhabiting 'foreign and cosmopolitan circles that are more various and accessible', this 'new bourgeoisie' had a lifestyle and morality that were very different to those of the 'traditional' French bourgeoisie (Halbwachs 1955, pp. 61–3). In *Distinction*, Bourdieu could be seen expanding on Halbwachs's insight, offering a detailed analysis of the new kinds of bourgeois and petty-bourgeois lifestyle which were the complements, on the cultural level, of the economic and social liberalism that characterised Giscard's 'progressive conservatism'.

The text of *Distinction* was illustrated with a montage of photographs and excerpts from magazines or newspapers, which were carefully chosen to highlight the confluence of political, cultural and economic phenomena which characterised the rise of the 'progressive fraction' of France's bourgeoisie. For example, Bourdieu juxtaposed a photograph of the high-tech interior of Servan-Schreiber's apartment with a photograph of a classically furnished bourgeois interior (1979, p. 359 [p. 313]); he included an article from *Madame Figaro* praising the taste in interior decor of Isabelle d'Ornano, sister-in-law of the Giscardian minister Michel d'Ornano (pp. 302–3 [pp. 268–9]), whilst Giscard's own delicate frame was juxtaposed with a photograph of a working-class bodybuilder (p. 233 [p. 210]). These excerpts from magazines and newspapers had been chosen not simply to illustrate the refined taste of the bourgeoisie per se, but rather to emphasise the new aesthetic and lifestyle pursued by the 'new' or 'progressive' bourgeoisie, as they struggled to impose a new definition of legitimate culture.

Strategies, Trajectories, Reconversions

According to Bourdieu, the rise of the 'progressive' fractions of the bourgeoisie, and of the political, cultural and ethical values they held dear, involved a challenge to the ascendancy of *belle lettriste* culture over more commercial forms of knowledge and aptitude. The rise of the progressive fractions of the bourgeosie thus heralded a shift in the amount of prestige attached to each of the forms of capital on offer in the social field. In a developed society with differentiated semi-autonomous fields, Bourdieu argued, a diverse range of forms of capital existed, educational capital, economic capital, cultural capital, even social capital, that network of contacts and influential friends so important to a successful career. Adherence to a class and, a fortiori, to a specific class fraction, depended not merely on the volume but also on the *structure* of capital possessed. Moreover, Bourdieu argued, in different historical conjunctures, different forms of capital would carry greater weight than others. At the heart of *Distinction*, then, was an attempt to trace shifts in the differential value attached to each form of capital and the consequent changes to the structure of capital possessed by the dominant social classes and fractions.

As Bourdieu argued, a three-dimensional map of social space could be constructed, 'whose three fundamental dimensions are defined by volume of capital, structure of capital and the evolution of these two properties over time' (1979, p. 128 [p. 114]). Individuals and groups inhabiting that social space were involved in 'strategies' to conserve or accumulate capital, 'reconverting' economic into educational capital, social into economic in accordance with historical circumstances. The existence of different forms of capital and differentiated 'fields' or 'markets' in which that capital could be invested meant that agents followed potentially divergent 'trajectories' through social space; trajectories whose outcome was not determined once and for all by social origin but which reflected the meeting of class habitus and 'the *field of possibles* objectively offered to a given agent' (p. 122 [p. 110]). The ability to select the correct trajectory to follow or the most profitable field in which to invest was less a matter of free choice than of *strategy*, that almost intuitive 'practical mastery' of the social field, which formed part of the bourgeois habitus.

Using correspondence analysis and drawing on statistical data concerning annual family income and profession (economic capital), highest educational qualification, profession of father and cultural tastes and practices (cultural capital), Bourdieu plotted the volume and structure of capital possessed by the different fractions of the bourgeoisie. He argued that cultural and economic capital were distributed 'chiasmically' within the dominant class. At the dominant pole of the dominant class were industrialists and senior executives who possessed high economic capital and relatively low amounts of cultural capital. At the dominated pole of the dominant class, this relationship was inverted and the intellectual fraction possessed high cultural capital and relatively low economic capital. Between these two fractions were the liberal professions who, in possessing almost equal amounts of economic and cultural capital, were the epitome of the cultured, wealthy bourgeoisie.

Each of these different fractions, Bourdieu argued, possessed their own characteristic ethos and hence quite different tastes and lifestyles. For example, he contrasted 'the ascetic aristocratism of the teachers', their preference for serious cultural pursuits, with 'the luxury tastes of the members of the liberal professions, who amass the most expensive and most prestigious activities' (p. 325 [p. 286]). Similarly, he maintained that while the intellectual fraction of the dominant class sought in art 'a symbolic challenging of social reality', the dominant fraction sought in high culture 'emblems of distinction which are at the same time means of denying social reality' (p. 334 [p. 293]). Clearly, Bourdieu was attempting to provide a more nuanced account of cultural capital, which, in his earlier work, had tended towards the monolithic and all-encompassing. Nonetheless, he remained doggedly sceptical as to the possibility of any cultural form offering a genuine challenge to the status quo. In opposing art to the base materialism of the market and the bourgeoisie, Bourdieu argued, the dominated intellectual fraction deployed the same oppositions between 'disinterest' and 'vulgar materialism' as the bourgeoisie as a whole employed to distinguish itself from the dominated classes. Hence, the most apparently radical art forms could be recuperated by the dominant fraction of the bourgeoisie for its own ends (p. 284 [p. 254]). Even avant-garde art, which apparently challenged the art institution itself, staged that challenge within the cultural domain and thus, 'by its very existence, helps to keep the cultural game functioning' (p. 280n.25 [p. 251]).

Having sketched out the distribution of the different forms of capital amongst the various fractions of the dominant class, Bourdieu sought to analyse the shifts in the differential value attributed to each of those forms of capital. Referring back to his earlier studies of higher education, *The Inheritors* and *Reproduction*, Bourdieu argued that the postwar 'schooling boom' was a key determinant in a series of changes to the relationship between the different classes, between those classes and the education system, and between educational qualifications and job prospects (p. 147 [p. 132]). These changes were provoking a realignment in the relationship between economic, educational and cultural capital. The influx of both women and the *classes moyennes* into higher education had greatly increased the number of university graduates. In order to retain the rarity value of their qualifications, Bourdieu argued, the dominant class had to invest more in educational capital. These developments merely produced 'inflation' in the educational market and a general 'devaluation' in educational qualifications. This devaluation had the paradoxical effect of increasing the importance of inherited social and cultural capital.

In a job market increasingly filled with qualified graduates, the importance of a network of personal or family acquaintances increased exponentially; social capital was one of the 'the means of resisting devaluation' (p. 149).[2] An equally important element of inherited cultural capital was that instinctive familiarity with the world of education, that highly refined *'sens du placement'*, the ability to spot a good investment combined with an almost intuitive sense of anticipation, which predisposed members of the dominant class to respond much more rapidly to changes in the educational and job markets, knowing 'the right moment to pull out of devalued disciplines and careers and to switch into those with a future' (p. 158 [p. 142]). Anticipating the devaluation of university degrees and the declining prestige of arts subjects, the dominant class abandoned the universities in favour of the more technocratically or commercially oriented education on offer in one of the selective *grandes écoles*.

Those who had recently gained access to higher education, women and the petty bourgeoisie, lacked this inherited '*sens du placement*' and continued, according to what Bourdieu termed 'the hysteresis of categories of perception and appreciation', to place their faith in those educational qualifications which had possessed value at an earlier historical moment. The resulting 'disparity between the aspirations that the educational system produces and the opportunities it really offers', had, according to Bourdieu, given birth to 'a cheated generation', amongst the petty bourgeoisie, forced, despite their qualifications, to accept relatively menial and uninspiring jobs (pp. 159–66 [pp. 143–7). Those with no formal qualifications were the most adversely affected by this devaluation of educational qualifications, finding it increasingly difficult to secure employment of any kind (p. 150 [p. 134]).

Bourdieu should be credited with considerable prescience here. For the trends he sketched were to continue throughout the 1980s and 1990s. The periodic violent protests of French students against any measure perceived as likely to erode either their access to higher education or the precarious privileges it guarantees might themselves be seen as a somewhat paradoxical response to the continued devaluation of French higher education degrees. Mass unemployment in France, meanwhile, means that it has become increasingly impossible to find the most menial of jobs without some higher or vocational qualification.[3]

These changes in the educational field coincided with and, to some extent, were determined by changes to the French economy, which, Bourdieu argued, was becoming increasingly driven by the demands of multinational business and the international banking system (p. 338 [p. 297]). In order to retain their wealth and power, the dominant class thus had to 'reconvert' its economic capital into educational capital, ensuring that its offspring, who could no longer depend on the family's 'rents', had the qualifications necessary to gain employment in multinational business:

> The reconversion of economic capital into educational capital is one of the strategies which enable the business bourgeoisie to maintain the position of some or all of its inheritors, by enabling them to extract some of the profits of industrial and commercial firms in the form of salaries, which are a more discreet – and no

doubt more reliable – mode of appropriation than 'unearned' investment income. (p. 155 [p. 137])

A new managerial class was emerging; 'executives in major national firms, public or private, ... or heads of large, modern, often multinational companies'; their position was justified by qualifications obtained from one of the rising band of elite business or administrative schools. Unlike their predecessors, they had no particular attachments to 'local privilege and prestige, which are increasingly devalued as the economic and symbolic markets are unified, setting them in the national or international hierarchy' (p. 360 [p. 314]). At an institutional level, the emergence of this new managerial class was mirrored in the rising fortunes of the more commercially or managerially oriented *grandes écoles*, such as the Institut d'études politiques (Sciences-po), the École des hautes études commerciales (HEC), or the École nationale d'administration (ENA), at the expense of the guardians of a more traditional or classically Republican culture and learning, the École normale supérieure (ENS) or the École polytechnique.

These shifts in the relationship between different forms of educational and economic capital were mirrored at the level of taste and lifestyle, Bourdieu argued, in the replacement of a certain classicism in bourgeois manners, interior decor, sport and leisure with a greater emphasis on high-tech interiors, a more relaxed, narcissistic morality of self-realisation, and the practice of what he termed 'Californian sports', archery, windsurfing, hang-gliding, and so on (pp. 241–6 [pp. 218–25]). Even those sports traditionally associated with the aristocracy or haute bourgeoisie, yachting, horse riding, or flying, were given a new valency when practised by 'a free-trading, multinational bourgeosie which derives its power from its decision-making and organisational capacities' (p. 241 [p. 218]). Forming what Bourdieu termed 'the new bourgeoisie', this new managerial class constituted a 'new ethical avant-garde' with its own specific ethos and lifestyle quite different from France's more traditional haute bourgeoisie:

> imbued with the economic-political culture taught in the political science institutes or business schools and with the modernistic economic and social world view which is bound up with it and which they help to produce in their conferences, commissions and seminars, these 'dynamic executives' have abandoned the champagne of the *vieille France* industrialists ... for the whisky of American-style managers and renounced the cult of *belles lettres* (delegated to their wives) for economic news, which they read in English. Being both the negation and the future of the old-style *patrons*, of whom they are often the heirs ... , they are

the ones who transcend the better to conserve. (pp. 360–1 [pp. 314–15, trans. modified])

This 'new bourgeoisie', Bourdieu maintained, was playing a central role in the imposition of a new ethos based on consumption rather than production; 'a hedonistic morality of consumption based on credit, spending and enjoyment [*la jouissance*]'. Employed in marketing, public relations and advertising, 'the new bourgeoisie of the vendors of symbolic goods and services' was responding to the demands of the 'new' economy by promoting consumer demand for a whole range of products and imposing a new set of prescriptions about taste and lifestyle (p. 356 [p. 310]). Bourdieu argued that this new emphasis on consumption had favoured the dominant class since those of its members who lacked the requisite educational capital could exploit their inherited cultural and social capital in the new professions in the media, marketing and advertising which had less rigidly defined entrance criteria and thus fulfilled the same role as had the army, the priesthood or the colonies in an earlier age (p. 168n.37 [p. 573n.27]). Indeed, Bourdieu maintained, the increased importance of financial and commercial functions over technical or scientific ones, under the impetus of an increasingly multinational economy, itself reflected in the rise of HEC, ENA, and Sciences-po, had enabled the haute bourgeoisie to strengthen its grip on power:

> Thus, as a result of changes in the economic structures, and chiefly through its use of the Paris Insitut de sciences politiques (Sciences-po), situated at the bottom of the properly academic hierarchy of the 'schools of power', the Parisian *grande bourgeoisie* has reappropriated, perhaps more completely than ever, the commanding positions in the economy and the civil service. (pp. 338–9 [p. 297])

Many of the features of Bourdieu's description of the bourgeoisie were to be reflected in his analysis of the petty bourgeoisie, both in terms of the 'chiasmic' distribution of forms of capital and the shifts in economic, educational and cultural fields. The petty bourgeoisie, he argued, was divided between artisans and shopkeepers, endowed with high economic capital but low cultural capital, and the primary teachers and 'new cultural intermediaries', whose low economic capital was compensated by their relatively high cultural capital (p. 395 [p. 342]). In between the two was 'the executant petty bourgeoisie', comprising junior executives and office workers.

The 'new cultural intermediaries', including all those who worked in 'occupations involving presentation and representation', from 'marriage counsellors, sex therapists, and dieticians' to 'youth leaders, tutors, monitors and TV producers and presenters', formed part of the most dynamic fraction, 'the new petty bourgeoisie' (p. 415 [p. 359]). The composition of this fraction reflected the expansion of higher education and the feminisation of the workforce, whilst its emergence as a social force was determined by changes in the economy, 'in particular, the increasing role of the symbolic work of producing needs, even in the production of goods' (p. 397 [p. 345]). It was in these new professions, associated with the production of the *symbolic* value of goods or services, that women were able to find a market for what were traditionally considered to be feminine qualities and aptitudes, their 'natural' charm, sociability, even physical beauty. Beauty, Bourdieu maintained, was acquiring 'a value on the labour market' and a 'redefinition of the legitimate image of femininity' was taking place under the combined effects of advertising, changing fashions, women's magazines and the emergence of a battery of advisors and educators, doctors and dieticians imposing new norms of body weight, diet and appearance through 'new uses of the body and a new bodily hexis', in the sauna, the gym and on the ski slope. These transformations substituted 'seduction for repression, public relations for policing, advertising for authority, the velvet glove for the iron fist', achieving 'the symbolic integration of the dominated classes by imposing needs rather than inculcating norms' (pp. 169–72 [pp. 151–4]).

Here, then, was an example of what Bourdieu identified in *The Logic of Practice* as the increasing reliance of developed Western societies on symbolic rather than straightforwardly coercive modes of domination. If this had particular implications for women, those implications were not limited to questions of gender politics. The 'new petty bourgeoisie' as a whole had abandoned the 'somewhat morose asceticism' of the traditional petty bourgeoisie to become the natural ally of the 'new bourgeoisie' (p. 423 [p. 366]). This alliance between the new petty bourgeoisie, the aspirational *classes moyennes*, and the dynamic new bourgeoisie was, as Jean-Jacques Becker (1998, p. 104) has shown, an integral part of Giscard's vision of a dynamic, libertarian and classless French society. For Bourdieu, the new petty bourgeoisie, having replaced the traditional petty-bourgeois respect for social norms and rigid hierarchies with an emphasis on the modern and the dynamic, found itself in the vanguard of 'the imposition of new doctrines of ethical salvation', based around the expression and realisation of *personal* needs and desires:

> Thus, whereas the old morality of duty, based on the opposition between pleasure and good, induces a generalised suspicion of the 'charming and attractive', a fear of pleasure and a relation to the body made up of 'reserve', 'modesty' and 'restraint', and associates every satisfaction of forbidden impulses with guilt, the

new ethical avant-garde urges a morality of pleasure as a duty. This doctrine makes it a failure, a threat to self-esteem not to 'have fun', or as Parisians like to say with a little shudder of audacity, *jouir* ... (p. 424 [p. 367])

In 'La Production de l'idéologie dominante', Bourdieu (1976, p. 44n.9) had argued that this emphasis on freedom of cultural and sexual expression had been born in May 1968 only to be appropriated and recuperated by the new bourgeoisie; it was 'one of those themes ... that the new discourse has retained from the discourse of May, whilst emptying it of any subversive force'. The apparent liberalisation of a once rigid sexual and social morality thus concealed, according to Bourdieu, an equally normative ethic, 'a cult of personal health and psychological therapy' whose injunctions to self-realisation were as prescriptive as the repressive morality they claimed to contest (p. 425 [p. 367]). This 'psychologisation of the relation to the body' was, he suggested, accompanied by the emergence of a market for a whole range of 'alternative' therapies, sporting and leisure practices: 'communes, creativity, dance, diet, drugs, ecology, encounters, esoterica ... Gestalt-therapy, ... grass, homeopathy, ... judo, kendo, kyudo, ... nomads, ... parapsychology, ... plants, pottery, prisons, psi phenomena', and so on (p. 429 [p. 370]).

Significantly, Bourdieu's list of alternative therapies and pastimes included some sly references to thinkers often considered 'postmodernists', to Deleuze and Guattari ('nomads') and to Foucault ('prisons'). His analysis of the new imperative to enjoyment, to '*jouir*' and '*jouissance*', might also be read as allusions to a certain discourse of sexual desire in contemporary French thought, in texts such as Roland Barthes's *The Pleasure of the Text* (1973) or Lyotard's *Libidinal Economy* (1974). In referring to 'the convergences between the routine themes of advertising – which has long been versed in the language of desire – and the most typical topics of philosophical popularisation', Bourdieu pointed to what he saw as the complicity between 'postmodern' thought and the new forms of consumerism (p. 431 [p. 371]). For Bourdieu, this new 'ethic of liberation' seemed to be 'in the process of supplying the economy with the perfect consumer', a consumer freed of all the old restraints of class and community; an isolated, atomised, depoliticised individual defined primarily by what they consumed (p. 431n.42 [p. 371]).

Bourdieu's critique of thinkers generally considered 'postmodernists' is significant since a series of commentators have interpreted his emphasis on the realm of consumption rather than production, on the body and its inscription by new discourses of consumer desire, as itself manifesting a characteristically 'postmodern' sensibility (Lash 1990; Featherstone 1991; Bauman 1992, p. 51). Thus, Chris

Wilkes has suggested an affinity between the role of the body in Bourdieu and its role in the work of Michel Foucault (in Harker et al. 1990, p. 117). Certainly, Foucault's description in *Discipline and Punish* (1975) of the shift in penal practices from a 'classical' regime of brutal physical punishments to 'modern' practices involving close surveillance, by a host of experts and advisors, of the criminal subject's psychological state, needs and motivations, did suggest one such affinity.

However, it is important to note the very particular conception of power, knowledge and subjectivity which informed Foucault's analysis. For Foucault, the category of the 'free subject' was precisely one product of the 'modern', of that shift from a regime of severe physical punishment to a 'scientific' discourse of criminality and penality. Having identified subjectivity as itself a construct of modernity, Foucault could no longer undertake a classically modernist critique which sought to rescue the essence of free subjectivity from its alienation by certain identifiable repressive discourses or material forces. If Foucault's thought has been designated 'postmodern', it is precisely because it demands we think beyond the modern category of the subject and all it entails.

Bourdieu's analysis, by contrast, continued to work within the terms of a classically modernist critique. If he portrayed the subject as being de-centred by a series of social and cultural forces inscribed on the body, he nonetheless considered that such forces could be identified and worked through in a process of rational sociological critique. As he put it in *The Logic of Practice*, sociology offered, 'perhaps the only means of contributing, if only through awareness of determinations, to the construction ... of something like a subject' (1980, p. 41 [p. 21]). Bourdieu's notion of the body thus signalled his adherence to a form of (modernist) critique whose ultimate goal was to liberate the subject through the rational grasp of those material determinants which threatened its freedom.[4] Such a project distinguished his work not only from Foucault's, but a fortiori from other 'postmodern' accounts of desire and the body such as that contained in Deleuze and Guattari's *Anti-Oedipus* or Lyotard's *Libidinal Economy*.

Similarly, Bourdieu's insistence that each class was defined 'by its consumption ... as much as by its position in the relations of production (even if it is true that the latter governs the former)' did not, in itself, make of him a 'postmodernist' thinker (1979, p. 546 [p. 483]). Indeed, such emphasis on the realm of consumption had an important precedent in the work of Halbwachs, particularly in his study of working-class lifestyles, *La Classe ouvrière et les niveaux de vie* (The Working Class and Standards of Living) (1912). Anticipating *Distinction* by over sixty years, Halbwachs had used data on working-class expenditure on food, clothing and housing to demonstrate that whilst position in the relations of production was important to class identity, so was consumption: 'if social distinctions do have their origin [in the realm of production], it is elsewhere, it is in society inasmuch as it does not produce but it consumes that those distinctions are most clearly manifested and delineated' (1912, p. 125).

As Baudrillard argued in *The Mirror of Production*, there was an entirely conventional notion of social distinction and of the way in which the consumption of certain commodities could signify wealth and prestige. Within such a conventional notion of distinction and consumption, distinctive value and the 'false' needs generated under consumer capitalism were still understood as somehow secondary or supplementary to a basic level of 'use values', of pure organic needs which existed prior to their alienation in the domain of capitalist exchange. If Baudrillard is often dubbed a 'postmodernist', it is because, in works such as *The Mirror of Production* and *For a Critique of the Political Economy of the Sign*, he asks us to think beyond such a conventional, modernist notion of consumption and distinction, to deconstruct the opposition between use and exchange value and the unstated opposition between nature and culture on which it relies. According to Baudrillard, all needs are always already culturally mediated; there is no domain of pure use value over against which exchange or distinctive value defines itself. Uncoupled from their articulation to any concept of use value, commodities are governed by an ungrounded play of signifiers: 'the signified and the referent are abolished in favour of the mere play of signifiers ... the use value of the sign disappears in favour of its simple value of commutation and exchange' (1973, p. 92).

Bourdieu's critique of consumerism and of the renewed emphasis it placed on symbolic distinctions, however, remained within an entirely conventional, modernist problematic. This renewed emphasis on distinction and consumption was understood in terms of the imposition of a series of supplementary, 'false' needs. Moreover, as David Swartz points out, this opposition between 'false' and 'real' needs, culture and nature, was nowhere clearer than in Bourdieu's portrayal of the French working class, whose lifestyle was seen as 'highly constrained by primary necessities', as though 'cultural practices become possible *only after* primary needs are satisified'; a view which rested on 'the classic opposition of nature and culture' (1997, p. 176).

The Working Class as 'Foil'

Bourdieu's portrayal of the working class in *Distinction*, more particularly his assertion that, 'there is no popular art' (p. 459

[p. 395]), has provoked harsh criticism even amongst his most favourable critics (Jenkins 1992, p. 148; Garnham in Calhoun et al. 1993, p. 181; Fowler 1997, p. 100). Central to Bourdieu's analysis of the aesthetic and lifestyle of the working class was the notion of the immediacy of material necessity. The precondition of the bourgeois aesthetic disposition was, he argued, that distance from the exigencies of material need which allowed the aesthete to suspend all question of the practical utility of a work of art and contemplate it on the level of pure form. The working class, on the other hand, because of the immediacy of their material needs, lacked the leisurely, contemplative distance on the world constitutive of the aesthetic disposition. Theirs was a realist aesthetic, an 'aesthetic "in itself" not "for itself"', which manifested itself in a whole series of functionalist preferences in interior decor, for quantity over quality in food, art, sport and so on (p. V [p. 4]). So strong was the class ethos based around this collective experience of necessity and 'virtue made of necessity', Bourdieu maintained, that even the relatively wealthy fractions of the working class retained a taste for the simple things in life for fear of earning the ridicule and disapprobation of their peers (pp. 452–8 [pp. 390–4]). Intellectuals who attributed an independent and potentially liberating popular culture to the working class were guilty of naive wish-fulfilment, of ignoring the material conditions which prevented a popular aesthetic from ever constituting itself.

Key to Bourdieu's argument were the conclusions he drew from a statistical analysis of the type and quantity of food consumed by working-class households. He argued that the preference of the working class for substantial rather than delicate or 'formal' cuisine, a manifestation of the immediacy of their material need, was paradigmatic of their whole experience of the world and particularly of their attitude to time. He contended that working-class eating habits manifested a demand for immediate satisfaction, an inability to defer or delay the moment of fulfilment in a rational calculation of possible future benefit:

> The 'modest taste' which can sacrifice immediate pleasures and desires to the satisfactions to come is opposed to the spontaneous materialism of the working classes, who refuse to participate in the Benthamite calculation of pleasures and pains, benefits and costs. In other words, these two relations to the 'fruits of the earth' are grounded in two dispositions toward the future which are themselves related in circular causality to two objective futures. (pp. 201–3 [p. 180, trans. modified])

Bourdieu's reference to 'Benthamite calculation' recalled Marx's critique of Jeremy Bentham's eagerness to endow every subject with a freedom of choice independent of their particular economic circumstances (Marx 1867, p. 602n.2). His attempt to deduce a specifically working-class ethos from a statistical analysis of the consumption of food echoed Halbwachs's *La Classe ouvrière et les niveaux de vie*. His distinction, finally, between two different dispositions towards the future, the one immediate, the other deferred and implying the capacity for rational calculation, reflected the influence of the now familiar Husserlian distinction between 'the forthcoming' [*l'à venir*] and 'the future' [*le futur*]. The working class, it was argued, had internalised their experience of material necessity into a class ethos based around the immediate satisfaction of present need, an essentially 'practical' habitus which lived for the day rather than investing hopes and energies in unrealisable projects for the future. This practical class ethos constituted a kind of immediate presence in the present:

> The being-in-the-present [*la présence au présent*] which is affirmed in the readiness to take advantage of the good times and take time as it comes is, in itself, an affirmation of solidarity with others … , inasmuch as this *temporal immanentism* is a recognition of the limits which define the condition. (1979, p. 203 [p. 183], my emphasis)

Just as the Algerian sub-proletariat had been judged incapable of constructing a rational political project for the future and the Kabyles incapable of achieving objective distance on their own socio-cultural practices, so the working class were taken to inhabit a realm of *doxa*, a kind of pre-reflexive immanence or absolute immediacy which prevented them from ever achieving the reflexive distance necessary to construct a genuine aesthetic. By a classically Romantic gesture, this apparently demeaning portrayal of the working class was used in *Distinction* as testimony to their authenticity, to an honesty and straightforwardness which Bourdieu contrasted with the 'calculated coldness' of bourgeois formality. As Swartz puts it, 'the French worker' was portrayed 'as a sort of "noble savage," freed from the invidious entrapments imposed by formalised culture' (1997, p. 176). Drawing a direct parallel between what he termed 'popular plain-eating' and 'popular plain-speaking', Bourdieu argued that in speech, as in food and art, the bourgeois tendency to formalise or euphemise was quite alien to the working class, who could not understand this 'sort of censorship of the expressive content which explodes in the expressiveness of popular language' (1979, pp. 35–6 [p. 34]).

Bourdieu's argument rested on his willingness to conflate the very distinct phenomena of *formalism* and *formality*, a conflation encouraged by the fact that the French word '*le formalisme*' covers both these meanings. For, if it is true that the working class manifests a greater willingness to employ words or expressions excluded from polite bourgeois conversation and that, as such, their speech betrays less *formality*, this does not necessarily equate to a simple expressiveness or lack of *formal complexity* in popular speech. On the contrary, popular speech forms, from Cockney rhyming slang through Glasgow 'patter' to the French *'verlan'*, manifest an extremely high level of formal complexity, often intended precisely to censor 'expressive content', albeit for rather different reasons than those of bourgeois propriety.

Similar objections could be raised to Bourdieu's analysis of working-class eating habits. For it is not necessary to deny the influence of economic factors in this realm in order to question whether the working-class diet does reflect, in any straightforward way, an immediate need for nourishment. As a comparative study of, for example, British and French working-class diets would demonstrate, at roughly comparable levels of income eating habits and quality of diet vary massively depending on the specificities of history, geography and national culture. In other words, the diet of the French working class does not reflect, in any mechanical way, the *immediacy* of their need for nourishment, but rather the whole range of historical, economic and cultural determinants through which that need is *mediated*. Once the necessarily mediated nature of working-class needs is taken into account, then the series of analogies which Bourdieu drew between the supposed immediacy of their material need, the 'expressiveness' of their speech, the inherently non-formal nature of their tastes and lifestyles, and their consequent inability to achieve the requisite measure of aesthetic distance to construct their own aesthetic, no longer holds.

If Bourdieu's analysis of the working-class aesthetic seemed to conflate formality with formalism, it also appeared to have confused questions regarding the *inherent capacity* of certain social groups for formal invention in the linguistic and cultural domains with consideration of the *institutional and societal constraints* which ensure that the richness and formal inventiveness of popular cultures are so frequently disparaged and denigrated. It is not necessary to deny marginalised groups' capacity for formal aesthetic invention in order to share Bourdieu's concern at the tendency of certain intellectuals to exaggerate the liberating potential of popular cultural forms.

Bourdieu's somewhat cursory treatment of working-class culture thus contrasted strongly with the highly nuanced and fundamentally dynamic vision of culture and cultural change to be found elsewhere in *Distinction*. Against the conventional wisdom which sees *Distinction* as fundamentally concerned with static cycles of social reproduction involving an unchanging model of 'legitimate culture', this chapter has highlighted the extent to which Bourdieu understood language and culture as contradictory entities, the sites of a constant struggle between classes and class fractions, subject to the pressures of significant social, economic and political change. Nonetheless, Bourdieu's tendency to understand bourgeois culture purely in terms of its distinctive value did pose a further problem. For the use of a single, all-encompassing term, namely 'cultural capital', to describe phenomena as diverse as the avant-garde theatre of Brecht, the classical culture of the Parisian haute bourgeoisie, and the newer managerial or neo-liberal culture, risked a certain relativism. In understanding the 'objective function' of all such cultural forms to be the legitimation of class hierarchy, Bourdieu seemed to be left with no grounds on which to defend the values, however compromised, of a 'humanist' culture against the incursion of market forces under the impetus of a full-blown neo-liberal ideology whose tentative beginnings *Distinction* had so presciently identified. The need to defend artistic and intellectual autonomy against just such forces was to become an increasingly pressing concern for Bourdieu throughout the 1980s and 1990s, both in his theoretical work and in his more directly political pronouncements. It is to a discussion of the problems raised by Bourdieu's attempts to conduct such a defence that the next chapter will now turn.

Notes to Chapter 6

1. This reference to Bakhtin's notion of 'double-voicedness' is by no means gratuitous. In *Language and Symbolic Power*, Bourdieu (1982, p. 18 [p. 40]) acknowledged the influence of M.M. Bakhtin/V. Voloshinov's study, *Marxism and the Philosophy of Language*, on his thinking about language, class and power.
2. Inexplicably, this passage does not appear in the English translation.
3. This was a theme Bourdieu was to pick up again in *The Weight of the World* (see Bourdieu 1993, pp. 605–20) [pp. 427–40]).
4. As Bourdieu put it in *Sociology in Question*: 'Bringing the tendential laws [of the social field] to light is a precondition for the success of actions aimed at frustrating them. ... [T]he dominated groups have an interest in the discovery of the law as such .... Knowledge of the law gives them a chance, a possibility of countering the effects of the law .... In short, just as it de-naturalises, so sociology defatalises' (1980a, p. 46 [p. 26]).

CHAPTER 7

Neo-Liberalism and the Defence of the 'Universal'

In *Distinction* Bourdieu had been concerned to trace the shift to a more individualistic, cosmopolitan, business-oriented culture in the lifestyles and tastes of France's bourgeoisie and petty bourgeoisie. This shift in the 'social field' as a whole, epitomised by the rise of the 'progressive fractions' of France's bourgeoisie, had also manifested itself in a series of homologous shifts in a range of other fields or sub-fields. Thus, the rise of the 'new bourgeoisie' and their allies, the 'new petty bourgeoisie', was mirrored in the political field in the rise of 'reconverted conservatism' as the new dominant ideology, in the educational field in the rise of the more technocratically oriented *grandes écoles* and business schools, and in the field of economic or industrial activity in the rise of financial and technocratic job functions in large multinational corporations.

Bourdieu has always been reluctant to specify the nature of the articulation between these different fields, to explain whether developments in one field were determined by or determining of changes in other fields, whether such developments were causally linked or simply correlated. He has drawn back, for example, from following Althusser and theorising the articulation between different fields in terms of determination 'in the last instance' by the economy, arguing that the relationship between fields can only be defined on a case by case basis, through empirical research (1992, pp. 84–5 [p. 109]). He prefers to speak of this relationship in terms of *homology*: for example, the relationship between the dominant fraction of the bourgeoisie, endowed with high economic capital and low cultural capital, and the dominated intellectual fraction, endowed with low economic capital and high cultural capital, has a homology, in the sub-field of higher education, in the relationship between academics in administrative posts (high bureaucratic capital, low intellectual capital) and academics who are experts in their discipline (low bureaucratic capital, high intellectual capital). Any change in the relative power of these different groups reflects a shift in the amount of prestige attached to the various forms of capital they possess, a shift, for example, in 'the rate of exchange' between cultural and economic capital.

Although rejecting the notion of last instance determination by the economy in each and every case, Bourdieu did suggest that the shifts he had begun to trace in *Distinction* were attributable to the strengthening of the economic or commercial pole of the social field and the consequent weakening of the properly cultural or intellectual pole. The 'rate of exchange' between economic and cultural capital had thus altered. For Bourdieu, fields function by analogy to magnetic force fields, with their poles of attraction and repulsion. Thus, as the economic pole became stronger, so its force of attraction increased and affected the whole of the field, drawing ever more individuals, groups and institutions into its ambit, like iron filings around a magnet. In his work of the 1980s and 1990s, Bourdieu has set out to trace the effects exerted by this strengthening of the economic pole in a series of other fields and sub-fields: education, artistic and intellectual production, politics and the civil service.

In *Distinction*, Bourdieu had already noted the effects of this strengthening of the economic pole in the field of French higher education, in the form of the rise of more technocratically and commercially oriented *grandes écoles* at the expense of the purveyors of a more classical knowledge and culture, the École normale supérieure and the École polytechnique. In *The State Nobility* (*La Noblesse d'état* 1989), Bourdieu was to offer a more detailed analysis of this trend and of its impact on the political and intellectual assumptions of the increasingly homogeneous business, political and administrative elite produced by France's *grandes écoles*. One of Bourdieu's central concerns in his contributions to the collaborative volume *The Weight of the World* (*La Misère du monde* 1993) was the way in which that same elite, imbued with a specifically French brand of neo-liberalism, was working to erode the rights and living conditions of some of France's most vulnerable social groups. Sections in both *Practical Reason* (*Raisons pratiques* 1994) and *Méditations pascaliennes* (1997) sought to provide a basis from which to defend 'universal' ethical standards of practice in politics and the civil service, against a background of high-profile cases of financial corruption amongst France's governing elite.

In the fields of intellectual and artistic production, Bourdieu has argued that economic forces have increasingly made themselves felt through the growing importance of the mass media, which threaten those fields' 'autonomy' by offering quick rewards of money and fame to the most compliant and fashionable intellectuals and artists. *The Rules of Art* (*Les Règles de l'art* 1992) was an account of the emergence of autonomous fields of literary and artistic production in late nineteenth-century France. An analysis of the *production* of autonomous cultural artefacts to complement *Distinction*'s account of their *reception*, the book was also an attempt to recapture a 'heroic' moment of artistic and intellectual autonomy at a time when Bourdieu considered that autonomy to be under threat from the combined forces of advertising, the media and the market. Two years later, in *Free Exchange* (*Libre-échange* 1994), Bourdieu published a series of conversations with the artist Hans Haacke, whose work both charts and challenges the negative influence of corporate sponsorship on artistic freedom. This critique of the damaging effects of advertising and the media on intellectual and artistic production was also at the heart of the pamphlet *On Television and Journalism* (*Sur la télévision* 1996).

As the shorter, more accessible format of *On Television and Journalism* made clear, Bourdieu's analysis and critique of the incursions of the market into the intellectual field aimed to spark a debate which extended beyond the boundaries of academia to reach the broader public.[1] From the late-1980s onwards, Bourdieu has increasingly sought to address this broader audience, frequently adopting the role of a public intellectual to intervene directly in contemporary political and cultural debates. In the late 1980s and early 1990s, he was involved in a public campaign aimed at securing the creative and intellectual autonomy of French state television by ending its dependence on advertising revenue (see Bourdieu 1988b; 1989a; 1990b). In 1989, he set up the review *Liber*: distributed internationally and translated into several languages, it was intended as an open forum for intellectual debate (see Collier 1993). In 1993, he was instrumental in launching 'a call for the establishment of an international parliament of writers'. The 'parliament' met in November of that year to discuss threats to intellectual autonomy as various as the *fatwah* against Salman Rushdie, the assassination of Algerian intellectuals, and the increasing influence of advertising and the media over the intellectual field (see Dutheil 1993).[2]

In a more recent pamphlet, *Acts of Resistance* (*Contre-feux* 1998), Bourdieu has collected a selection of his political speeches, all intended, to quote the subtitle of the French edition, 'to serve as resistance against the neo-liberal invasion'. *Acts of Resistance* contains evidence of Bourdieu's most high-profile political intervention to date, his vociferous public support for the striking French students and workers in the autumn of 1995 in their protests against the right-wing government's proposed package of reforms and budget cuts affecting the universities, the nationalised railways, health care, pension rights, and social security provision. Significantly, Bourdieu's support for the strikers brought him into conflict with his old rivals in the French New Left, since the government's intended reforms had received the support of Nicole Notat, leader of the CFDT trade union, and of a group intellectuals close to the journal *Esprit*.[3]

On one level, then, Bourdieu's recent theoretical works and his more directly political interventions can be seen as the logical culmination of criticisms of a technocratic discourse, and of the French New Left's complicity with that discourse, which he had first made in the 1960s. Bourdieu's recent theoretical works and his overtly political pronouncements betray a series of related concerns with the increasing power of the market, with the rise of neo-liberal ideologies in France and the wider world, and with the threats posed to artistic, intellectual and political autonomy by such developments. He has increasingly couched this critique of the unfettered power of the market in terms of a defence of the 'universal' values of art, reason and ethics (Bourdieu 1992a, pp. 461–72 [pp. 339–48]; 1997, pp. 111–51). This 'universal' is not, Bourdieu has argued, grounded in a concept of transcendental or atemporal value, but rather is seen as the by-product of a specific set of historical circumstances which oblige individuals and groups to work 'to advance the universal' even as they pursue their own partial interests.

Nonetheless, this defence of artistic and intellectual autonomy has compelled Bourdieu to acknowledge that such autonomy can possess a value not reducible to issues of class and social distinction. This would seem to mark a significant change of approach from *Distinction*, where, as several critics have noted, he appeared to reject outright the possibility that legitimate culture might have any inherent value over and above its implication in questions of status and social distinction (Wolff 1983, pp. 36–8; Wilson 1988; Bürger 1990). Having analysed Bourdieu's account of the emerging neo-liberal hegemony across the French educational, administrative, political, intellectual, cultural and social fields, this chapter will thus examine his efforts to reconcile his recent defence of 'the universal' with an earlier body of work that seemed to reject any such claim to the universal as simply so much bourgeois bad faith.

The Grandes Écoles *and the Production of a 'State Nobility'*

Before considering Bourdieu's analysis of the production of the French business, administrative and political elite in *The State Nobility*, it is important to understand something both of the role of France's highly selective *grandes écoles* in this process and of the ways that role has evolved in the postwar period. The most prestigious *grandes écoles* have direct links with the French civil service, guaranteeing their best graduates posts in its upper echelons. For example, at the Liberation, the École nationale d'administration, (ENA), was established precisely to provide a 'Republican elite' of highly trained state bureaucrats, selected purely on merit, to oversee France's postwar reconstruction. If this alone gave graduates of ENA considerable power over the French state bureaucracy, a series of subsequent developments in postwar France has enabled them to extend that power into the domains of politics and business.

As Anne Stevens argues, the fact that the Fifth Republic constitution was drawn up in 1958 by the founder of ENA, Michel Debré, with the specific aim of increasing the powers of the executive, has helped graduates from ENA to consolidate their dominance over French politics and the civil service (1981, p. 139). According to Pierre Birnbaum, this 'fusion of political and administrative powers' under the Fifth Republic was given a new twist during Giscard's presidency, which was marked by an unprecedented 'osmosis' between the upper echelons of the French civil service and the business world (1994, p. 160). Moreover, the influence of high-ranking civil servants over the business world has been facilitated in postwar France by the existence of significant numbers of wholly or partially state-owned industries. In short, the peculiarities of the French *grande école* system mean that the best graduates from these highly selective higher education institutions are not only guaranteed posts in the *Grands Corps* at the summit of the civil service, but also have access to the worlds of politics, through sitting on one of the various *cabinets ministériels*, and of business, through supervision of one of France's major publicly-owned or private–public commercial enterprises.

It is partly this tradition of economic *dirigisme* which explains why France has never undergone a neo-liberal revolution directly comparable to the advent of Thatcherism in the United Kingdom or Reaganomics in the United States. Despite relatively modest privatisation programmes pursued by successive governments after 1986, France has yet to experience an all-out assault on the public sector on the Anglo-Saxon model. As Emmanuel Godin puts it; 'in France, neo-liberalism has provided a starting point for rethinking the *modalities* of state intervention, rather than for challenging its legitimacy' (1996, p. 67). France's ruling elite has thus not so much abandoned all faith in state intervention, as reinterpreted the discourse, goals and methods of neo-liberalism through the medium of a peculiarly French tradition of centralised state planning or *dirigisme*.

In *The State Nobility*, Bourdieu sought to analyse these trends, to trace the emergence of an increasingly socially homogenous ruling elite, imbued with a specifically French form of neo-liberalism. By following the career paths of former students of the *grandes écoles* into powerful positions in politics, business and the civil service, Bourdieu was able to examine the relationship between the field of higher education and what he termed 'the field of power' in far greater detail than he had before. Further, by using data from the mid-1960s to the mid-1980s, he could now trace the evolution of the field of French higher education over a much longer historical time span than had been the case in his earlier studies of the field.

Using correspondence analysis based on data for the geographical and social origins, political and religious affiliations, sporting, cultural and leisure activities of the student intake of 1966, Bourdieu plotted the state of the field of the *grandes écoles* at that time (1989, p. 222 [p. 156]). This would later serve as a benchmark against which the extent of the field's evolution over the intervening two decades could be assessed. He argued that in 1966 the field of the *grandes écoles* was structured 'chiasmically', according to a set of oppositions homologous to those he had identified in *Distinction* as dividing the different fractions of the dominant class.

At one pole of the field stood the École des hautes études commerciales (HEC), whose students were drawn primarily from the offspring of industrialists and high-ranking businessmen; they tended to be Catholic, have right-wing political affiliations, 'classical' tastes in culture, and practised sports such as rowing, tennis or golf.[4] Bourdieu described HEC, and institutions like it, as being 'heteronomous': closest to 'the field of power', the teaching they offered was most directly subordinate to the needs of business and was hence least concerned with transmitting 'specifically scientific' or 'exclusively academic' forms of knowledge.

At the opposing, more 'autonomous' pole of the field was the École normale supérieure (ENS), whose students were frequently the offspring of teachers or lecturers; they were more often atheist or agnostic, left-leaning with avant-garde cultural tastes and little interest in sport.[5] Unlike HEC, ENS enjoyed considerable 'autonomy' from the field of economic and administrative power and its teaching followed more 'properly academic' or 'scientific' principles. Between these two extremes stood schools such as the École polytechnique, or ENA, which recruited primarily amongst the sons and daughters of the liberal professions or civil servants.[6] Thus, HEC recruited its students primarily amongst the dominant fraction of the dominant class, which Bourdieu had described in *Distinction* as being characterised by its high economic capital and relatively low cultural capital. ENS, on the other hand, recruited principally from the dominated fraction of the dominant class, the intelligentsia, characterised by its high cultural capital and relatively low economic capital.

Tracing the evolution of the field of the *grandes écoles* from the 1960s through the 1970s and into the 1980s, Bourdieu concluded that those *grandes écoles* which recruited amongst the dominant fractions of the bourgeoisie had grown even more powerful. Graduates who had passed through HEC or Sciences-po, before completing their education at ENA, were increasingly monopolising positions of power in French politics, administration and industry, whilst the dogmas of managerialism and business efficiency they espoused were having a deleterious effect over the whole intellectual field. Comparing his data for entrants into the *grandes écoles* in 1966–67 with data for the academic year 1984–85, Bourdieu noted that the number of successful candidates from modest social backgrounds had actually declined over the period. This he attributed to the massive expansion in the university sector, which, contrasting strongly with little or no increase in the number of places available in the *grandes écoles*, had rendered the latter even more prestigious, selective and elitist.

The extension of university education to groups previously excluded, notably women and the petty bourgeoisie, had been accompanied by a decline in the rarity value of university degrees, Bourdieu argued. In the wake of the events of 1968, particularly, the bourgeoisie had become increasingly determined that its offspring should not be educated in a university sector associated with such devalued qualifications, with overcrowding, student unrest and political radicalism. The development of this 'new educative demand' had combined with changes to the job market, contingent on 'transformations in the economic field such as the growth in international trade', to encourage the emergence throughout the 1970s of a mass of new business and management schools, 'a nebula of academic institutions more directly geared to the demands of the corporate world' (pp. 305–28 [pp. 214–29]).

If this combination of causal factors in both the educational and the economic fields had given birth to a mass of new business and management schools, it had also impacted on the traditional hierarchy of academic prestige between the older *grandes écoles*. As the economic field became increasingly global, so businesses became ever more dependent on international finance capital and the greater control exerted by banks over whole sections of industry was reflected in a change in the balance of power between different management functions; 'a strengthening of financial management in relation to technical management, in other words of Inspectors of Finance and Sciences-po graduates in relation to those holding technical diplomas such as Mines engineers' (p. 468 [p. 327]).[7]

According to Bourdieu, *grandes écoles* such as ENA, which trained the Inspectors of Finance, and Sciences-po were less 'autonomous' and their academic qualifications less 'specifically scientific' or 'exclusively academic' than 'engineering schools' such as the École polytechnique or the École des Mines (p. 215 [pp. 153–4]). Such schools demanded relatively little of their students in terms of academic ability and equipped them with a set of general 'technocratic' or 'bureaucratic' aptitudes, which Bourdieu contrasted with the more rigorous 'technical' training on offer at Mines or the

École polytechnique. He argued that the entrance examination to ENA represented the apogee of a test which rewarded 'general culture' and rhetorical ability over genuine knowledge: a test destined, therefore, to favour those endowed with the habitus of the cultured Parisian haute bourgeoisie. ENA was the epitome of 'a sanctuary school', of which the numerous new management schools were also examples, in that it allowed the less academically gifted offspring of the dominant class to adopt a 'roundabout route' to economic and political power. However, in the changed economic and educational conjuncture of the 1970s and 1980s, it was precisely these 'sanctuary schools' which had gained prestige at the expense of the more autonomous, academically rigorous *grandes écoles*. The 'decline of technical qualifications to the advantage of qualifications guaranteeing a general culture of a bureaucratic kind', had, thus, benefited not only more 'heteronomous' institutions such as ENA, HEC and Sciences-po, but also the sons of the Parisian haute bourgeoisie amongst whom they traditionally recruited (p. 386 [p. 272, trans. modified]).

Bourdieu argued that even more traditional, provincial industrial dynasties, typified perhaps by Michelin in Clermont-Ferrand or the large textile firms of northern France, which would once have avoided the state education system, relying instead on private Catholic colleges and passing jobs down through the family, were coming to demand higher qualifications of all their executives. A 'family mode of reproduction' was being supplanted by 'a school-mediated mode of reproduction', and, as a general rule, the older and bigger a firm, the more its management had recourse to 'a form of academic consecration'. This was not to say that 'a mode of domination based on property and *owners*' had given way to 'another more rational and more democratic one, based on "competence" and *managers*' (p. 428 [p. 300]). Not only were the kinds of aptitude demanded of managers in these 'large companies under technocratic control' entirely consonant with the bourgeois habitus but, argued Bourdieu, under the 'school-mediated mode of reproduction', social capital, sustained by a network of family members in powerful positions, took on an added importance. Where previously family members had been rivals competing for an inheritance, now they could become allies sustaining a collectively held stock of social, educational and cultural capital (p. 416 [p. 292]).

Moreover, this shift in the 'mode of reproduction' was mirrored in the changing nature of bourgeois matrimonial strategies, with parents abandoning 'their *dirigiste* policy' as regards their children's choice of marriage partner. The new educational and economic conjuncture had completely redefined the criteria determining 'the value of their daughters on the matrimonial market', replacing the old emphasis on economic capital in the form of a dowry or on 'symbolic capital in respectability (virginity, appearance, etc.)' with an emphasis on educational or cultural capital. The greater freedom accorded the sons and daughters of the bourgeoisie in their choice of partner concealed the fact that the massive entrance of women into higher education provided an opportunity to meet a socially homogeneous group of possible future partners, hence ensuring 'homogamy at least as effectively, but much more discreetly, than by family interventionism' (p. 390 [p. 275]).

Bourdieu was playing with the terms *'dirigisme'* and 'interventionism', here, to draw a parallel between social or sexual liberalisation and the economic liberalism espoused by Giscard. He went on to describe the reforms introduced by Giscard in the domain of divorce, marital law and abortion as 'the political accompaniment, necessary for adjusting norms to practices, of a transformation of the mode of reproduction in force in the upper bourgeoisie' (pp. 390–1 [p. 275]). Economic liberalisation, the liberalisation of social and sexual mores, as well as shifts in the balance of power within the field of higher education, were thus all seen as part of a related process, a process, moreover, which greatly benefited the Parisian haute bourgeoisie. For, Bourdieu argued, it was this social group which was best placed to take advantage of the postwar economic and educational conjuncture. Not only was their habitus particularly well adjusted to the aptitudes demanded by 'the schools of power' , ENA, HEC, Sciences-po, but they were closer, both geographically and psychologically, to those institutions than the more traditional, provincial class of industrialists:

> Thus these categories, distinguished by a more open relationship to the social world, found themselves in a much better position than the great Catholic provincial bourgeoisie to benefit from the opportunities for ascension and reconversion offered by the new mode of reproduction and the new access route to positions of power opened up by the *grandes écoles* and especially, after the Second World War, by the École nationale d'administration. (pp. 414–15 [p. 290])

ENA had now become the most prestigious of France's *grandes écoles*, definitively displacing ENS, which in a previous age, notably under the Third Republic (1875–1940), had produced France's most influential intellectuals, such as Bergson and Sartre, and important politicians, such as Léon Blum and Jean Jaurès. ENA's predominance was manifest in the fact that graduates of ENS were now increasingly seeking to gain access to the school and complete their higher education there. If ENA's dominance in the educational field was manifest in the influence it was able to exert on graduates of the once dominant ENS, the nature of ENA itself had changed: it had been drawn towards the field's heteronomous pole to become ever closer to commercially oriented schools such as HEC or L'École Centrale des arts et manufactures (Centrale). Bourdieu bemoaned this 'curious drift' of ENA away from its original democratic, meritocratic vocation and public service ethos towards commercial values which mirrored those of its increasingly haut bourgeois clientele:

> the curious drift of an institution that – originating out of a stated and without any doubt sincere intention to rationalise and democratise the recruitment of higher public servants by doing away with dynasties founded on nepotism and the insidious heredity of offices – has come to fulfil a function altogether similar to the one that fell to HEC and Centrale at the end of the nineteenth century, that is, providing the bourgeois children socially predestined to occupy dominant positions with the academic guarantee that the most academically legitimate institutions were increasingly refusing them during a period of intensified academic competition. (p. 283 [p. 200, trans. modified])

The 'curious drift' of the ENA away from its intended function was mirrored, Bourdieu argued, in 'the field of power' itself in the increasing accommodation of the French civil service as a whole to the interests of business and the market. There was a relation of homology and 'causal interdependence' between the field of higher education and the field of power (p. 373 [p. 263]). The distribution of economic and cultural capital in the latter was arranged chiasmically, with business leaders endowed with high economic capital and relatively low cultural capital at its dominant pole, the intelligentsia, endowed with high cultural capital and relatively low economic capital, at its dominated pole. At the centre of this chiasmus were the upper echelons of the civil service (p. 382 [p. 270]). However, just as ENA had undergone a 'drift', slipping from its public service ethos to ever greater accommodation to directly economic forces, so the upper echelons of the civil service increasingly served as the point of exchange between administrative, political and economic interests. The existence in France of huge wholly or partly state-owned enterprises not only encouraged the phenomenon of *'pantouflage'*, whereby civil servants left public service for lucrative posts in industry and commerce, it also provided a peculiarly centralised arena of political, economic and administrative power. The Parisian bourgeoisie was uniquely equipped to profit in this arena, exploiting its networks of contacts, its inherited social and cultural capital, as well as the educational capital it had accumulated in the *grandes écoles*.

At the nexus of these different forms of power, administrative, economic, political, were those Bourdieu termed *'les grands patrons d'État'*, 'the top state chief executives'. Frequently of Parisian haute-bourgeois origin, educated in a prestigious Parisian *lycée*, a *grande école*, and ENA, they had typically passed through one of the *grands corps*, as well as a *cabinet ministériel*, to now enjoy an unrivalled amount of power and influence over the economic, political and social fields:

> The top state chief executives ... and also, but to a lesser degree, the old fractions of the bourgeoisie (officers or landowners), were 'predestined', as it were, to occupy positions located at the *intersection* between the public and the private sectors, or, better still, between banking, industry and the state, the very locus of power today. Indeed, everything prepares these 'men of connections' to occupy these eminent positions (top merchant banks, public energy and transport companies, public-private companies, etc.), where, in an atmosphere of complicity and conflict, large state markets and subsidies to so-called 'grass-roots' or 'high-tech' industries are negotiated and where political decisions (on credit, housing, etc.) likely to provide new areas of investment and new sources of profit are hammered out. (p. 472 [p. 329])

Bourdieu's concern was not merely with the socially homogeneous nature of this business and administrative elite, but also with the way in which, moving between the public and private sectors, they brought into the heart of the civil service the dogmas of business efficiency and the market which struck at the heart of a public service ethos.

In *Distinction*, Bourdieu had sought to trace the impact of these 'progressive' fractions of the bourgeoisie primarily in the domains of culture and lifestyle. *The State Nobility* provided an opportunity to look in greater detail not only on the role of the *grandes écoles* in producing and reproducing this rising class, but also at the political effects of their dominance over politics, business and the civil service, over France's 'field of power'. However, if Bourdieu's point about the dominance of this socially, culturally and ideologically homogeneous elite was well made and amply supported by detailed statistical analysis, his definition of 'the field of power' remained somewhat vague. His description of the field of power suggested it was merely an all-encompassing arena defined by the balance of power or rate of exchange between the different forms of capital at any given historical moment: 'a field of forces defined by the state of the relations of power between the different forms of power or the different forms of capital' (p. 375 [p. 264]). However, if the field of power amounted to nothing more than the aggregation of forms of capital accumulated in the various sub-fields, there seemed to be nothing to distinguish it from the social field as a whole.

When Bourdieu attempted to map the field diagrammatically, he did so by using correspondence analysis to plot the social space occupied by a selection of professional groups, from artists and university lecturers through the liberal professions to business leaders. Yet he omitted to plot the position of professional politicians as a separate group, to say nothing of the various legislative and executive institutions in which those politicians worked and which, in theory at least, are intended to represent and enact the will of the majority as expressed in democratic elections. In short, power is surely more than the sum of the forms of economic, cultural or educational capital possessed by a nationally defined selection of professionals. Moreover, Bourdieu's diagrammatic representations of the field of power seemed unable to account for the relationship between national institutions or nationally defined forms of power and those increasingly important supra-national institutions and forms of power, represented by multinational corporations or organisations such as the European Union and the World Trade Organisation. His emphasis on the homologies between the different fields, finally, seemed to offer a too-symmetrical view of the social world on the model of a Russian doll, each sub-field simply representing a mirror image in miniature of the larger field which encompassed it.

Where Bourdieu's analysis of the field of power was more sure-footed was in his use of a combination of biographical detail, statistical data and ethnographic description to trace the shared background, education, culture and networks of professional and social activity that characterised the French 'state nobility'. Details of educational trajectories, of memberships of company boards or *cabinets ministériels*, and statistical data regarding the prevalence of '*pantouflage*', or the number of political leaders of almost all shades, from the Gaullists to the Socialists, who had passed through ENA, all contributed to a convincing portrait of 'the evolution towards self-enclosure of a corps whose social triumph tends to make it all the more certain of itself because ... nothing comes to disturb its "splendid isolation"' (1989, p. 369 [p. 260]).

If there was one individual, according to Bourdieu, who personified the ethos of the French state nobility, it was Giscard d'Estaing. Giscard's trajectory, 'from the most traditional fractions, close to Pétain' of the French bourgeoisie, through the École polytechnique, ENA, the Inspection des Finances and the *cabinet ministériel* of Edgar Faure, to the political avant-garde of the 'new bourgeoisie', typified 'the entire evolution of the bourgeoisie' in postwar France (1989, p. 390 [p. 275]). In *The State Nobility*, as in later works, Bourdieu identified Giscard's presidency as marking a key moment in the erosion of the French civil service's public service ethos in favour of more commercial values.

While Giscardian liberalism may appear to have been convincingly defeated by the election of François Mitterrand in 1981, as Bourdieu was to remark in *The Weight of the World*, from 1983 onwards the Socialists themselves adopted a discourse and a set of policies which demanded submission to the 'realities' of the market (1993, pp. 220–1 [p. 182]). This 'rallying of the Socialist leadership' to 'the neo-liberal vision' after 1983 was symbolised by the replacement of prime minister Pierre Mauroy, an old-style Socialist politician with a power base in the traditional working-class, industrial city of Lille, by Laurent Fabius, a young graduate of ENA, and hence one further personification of Bourdieu's 'state nobility'.

In *The Weight of the World* as in *The State Nobility*, Bourdieu would charge France's governing elite, civil servants and politicians alike, with having 'abdicated' their democratic, republican responsibilities. Indeed, Bourdieu's critique of the 'curious drift', which had led ENA to abandon its vocation to 'democratise and rationalise' the recruitment of France's highest civil servants, had eloquently communicated his own sense of disappointment that this peculiarly French republican ideal had been betrayed. The sense that Bourdieu still had some residual faith in French republican ideals seemed to find confirmation in his distinction between 'autonomous', 'properly academic' institutions, such as ENS, and 'heteronomous', technocratically oriented schools, such as ENA.

For, although he criticised the elitism of the *grandes écoles* system as a whole, Bourdieu did argue that the general technocratic aptitudes rewarded by the least autonomous *grandes écoles* were particularly closely attuned to the habitus of the Parisian bourgeoisie. Further, the general nature of the qualifications awarded by institutions like ENA suited a job market in which the rise of general management or 'technocratic' job functions had blurred the relationship between the formal qualifications one possessed and the occupations those qualifications allowed one to perform. Where this blurring favoured the bourgeoisie, rewarding their 'general culture' and social capital, the dominated classes had 'in global terms, an interest in the rationalisation of both required skills and the manner in which they are taught and evaluated, as well as that of the definition of the attributes guaranteed by qualifications' (p. 170n.8 [p. 413n.10]). Bourdieu's emphasis on the desirability of rationalising qualifications and their relationship to the occupations one could rightfully occupy suggested a residual faith in the Durkheimian and typically republican ideal of 'organic solidarity'; a vision of a meritocratic society in which individuals would occupy positions in direct accordance with their inherent abilities, as attested to by the possession of the requisite qualifications.

Meanwhile, in introducing distinctions between 'autonomous' and 'heteronomous' educational institutions, dispensing more or less 'properly academic' forms of knowledge, Bourdieu had prepared the ground for the defence of intellectual autonomy and the critique of the free market ideologies of the 'state nobility' which were increasingly to characterise his output.

The Weight of the World *and 'The Abdication of the State'*

Where *The State Nobility* had attempted to trace the emergence of a political, administrative and business elite, imbued with a specifically French form of neo-liberalism, *The Weight of the World* constituted a cry of protest against the social effects of the policies pursued by that elite. The work of more than twenty sociologists, published under Bourdieu's general editorship, *The Weight of the World* took the form of transcriptions of interviews, prefaced by short explanatory commentaries, with individuals deemed to be suffering in some way from the effects of a neo-liberal consensus promoted by a ruling elite ever more divorced from the social and economic problems faced by significant sections of society. The interviews, Bourdieu argued, took the form of a kind of Socratic dialogue in which the sociologist merely enabled the research subjects to 'to bear witness, to make themselves heard, to bring their experience from the private into the public sphere' (1993, p. 915 [p. 627]). This 'form of maieutic' was opposed not only to the neo-liberal dogmas of the ruling elite but also to the hasty, sensationalist treatment social problems so often received in the media (p. 917 [p. 619]).

The scope of *The Weight of the World*, with its inclusion of sections on the ghettos in American cities, clearly extended beyond the French experience. However, Bourdieu's own contributions, whether individually or jointly authored, concerned France alone, focusing on problems of industrial decline, youth unemployment and the problems of France's multi-ethnic city suburbs. He argued that such problems were greatly exacerbated by the gulf that separated the political and administrative elite from large sections of the population. The world of politics was becoming ever more self-regarding, 'increasingly closed in on itself, on its internal rivalries, its own problems and issues' (p. 941 [p. 627]). Politicians were less interested in expressing the needs and aspirations of their electors than in satisfying 'the superficial demand to ensure their own success, making of politics a thinly-disguised form of marketing' (p. 942 [p. 628]). Journalists and intellectuals were similarly beholden to the demands of the market. This political vacuum, marked by the absence of any party or institution capable of representing the needs of the least powerful in society, could all too easily be filled by the populist demagoguery of the *Front national*: 'all the signs of all kinds of malaise are there and, for lack of being legitimately expressed in the political domain, they sometimes express themselves in the absurdities of xenophobia and racism' (p. 941 [p. 627]).

Returning to the Béarn to document the death throes of a peasant way of life whose decline he had first charted in 1962 in 'Célibat et condition paysanne', Bourdieu interviewed two peasant farmers who somewhat shamefacedly admitted to a sympathy for the *Front national*, the only political party to show any interest in their plight (pp. 519–31 [pp. 381–91]). An interview with local activists for the *Parti socialiste*, meanwhile, revealed an ill-concealed disgust at a leadership increasingly divorced from and indifferent to the concerns of its grass-roots membership (pp. 433–45).[8] However, it was in a section entitled, 'The Abdication of the State', that Bourdieu made it clear whom he considered to carry most responsibility for France's current social, economic and political problems. As its title suggested, this section took the French state to task for abdicating its responsibilities towards its citizens. More specifically, Bourdieu attempted to identify the housing policies responsible for the state of France's notorious *'banlieues'* (city suburbs), widely stigmatised as places characterised by racial tension, crime and social unrest. He argued that the policies pursued by Giscard's housing minister, Jacques Barrot, which had encouraged private property ownership and shifted emphasis away from state financing of physical structures towards giving aid to individuals, marked the beginnings of a process of liberalisation whose effects were now becoming all too clear:

> it is the withdrawal of the state and the decline of public subsidies for building, affirmed in the course of the 1970s in the replacement of subsidies for physical structures with subsidies for individuals, which is largely responsible for the appearance of sink estates where, under the effect of the economic crisis and unemployment, the most deprived populations are concentrated. (p. 220 [p. 182])

In *The State Nobility*, Bourdieu (1989, p. 436n.8 [p. 440n.8]) had identified the new housing policies introduced in the 1970s as a prime example of a policy change which encouraged the 'interpenetration' of high-ranking civil servants, politicians and businessmen; a point he and his collaborators had analysed in greater detail in a special number of *Actes de la recherche en sciences sociales* dedicated to the housing market (Bourdieu 1990). The housing policies of the Giscardian presidency were thus symptomatic for Bourdieu of a more general move away from the belief in state intervention in the name of the collective good of society towards a liberal or neo-liberal vision of citizens as individual, atomised consumers, recipients, as he put it in *The Weight of the World*, of 'a *state charity*, destined, as in the good old days of religious philanthropy, for the "deserving poor"' (p. 223 [p. 184]). This 'transformation of the (potentially) mobilised people into an aggregate of the atomised poor, of the "marginalised"' (p. 223 [p. 184]), reflected what Bourdieu termed, 'the collective conversion to the neo-liberal vision which, begun in the 1970s, was completed in the middle of the 1980s, with the rallying of the Socialist leadership' (pp. 220–1 [p. 182]).

Not only had mainstream left- and right-wing parties been converted to this neo-liberal dogma, he argued. It also expressed, 'very directly, the vision and the interests of the great state nobility, graduates of ENA and trained in the teachings of Sciences-po' (p. 221 [p. 183]). These 'new mandarins', secure in their guaranteed posts at the summit of the civil service, 'have pretensions to manage public services like private businesses, whilst all the while sheltering from the constraints and risks, financial or personal, associated with institutions whose (bad) morals they ape' (p. 222 [p. 183]).

This 'demolition of the idea of public service', meant that the state had adopted an ideology which ran counter to its declared aims and intentions, even to its very *raison d'être* (p. 221 [p. 182]); a contradiction which was experienced most acutely by the 'street-level bureaucracy' at the point of delivery of aid and social services. Bourdieu presented a series of interviews with these 'street-level bureaucrats', a social worker, a drugs outreach worker, and a probation officer whose efforts received little validation or official backing from the authorities who nonetheless expected them to resolve ever more pressing social problems (pp. 229–56 [pp. 189–212]). As a solution, Bourdieu proposed that the state should recognise its responsibilities and he advocated the adoption of 'a resolute policy by a state determined really to employ the means necessary for achieving its declared intentions' (p. 228 [p. 188]).

'The Corporatism of the Universal'

In a chapter of *Practical Reason* entitled 'Rethinking the State: Structure and Genesis of the Bureaucratic Field', Bourdieu offered a more detailed theorisation of his conception of the state, its emergence, powers and functions. Inflecting slightly Weber's (1968, p. 54) definition of the state as 'a compulsory political organisation' whose 'administrative staff successfully upholds the claim to the *monopoly* of the *legitimate* use of physical force in the enforcement of its order', Bourdieu defined the state as 'an X (to be determined) which successfully claims the monopoly of the use of physical and *symbolic* violence over a defined territory and over the totality of the corresponding population' (1994a, p. 107 [p. 40]). In brief, he argued that the state was the result of the concentration of various forms of capital, 'capital of physical force or instruments of coercion (army, police), economic capital or, better, informational capital, and symbolic capital'. This process of concentration led to the emergence of a specific form of capital, 'properly statist capital (*capital étatique*) which enables the state to exercise power over the different fields and over the different particular species of capital and especially over the rates of exchange between them' (1994a, p. 109 [pp. 41–2]).

This 'statist capital' was not, however, purely coercive for, Bourdieu argued, in delegating power to state institutions or personnel, an absolute ruler necessarily endowed the 'bureaucratic field' with a certain limited autonomy. In order both to legitimate and strengthen this autonomy, agents in the bureaucratic field had an interest in giving:

> a universal form to the expression of their particular interests, in elaborating a theory of public service and public order, and thus in working to autonomise the reason of state from dynastic reason, from the 'house of the king', and to invent thereby the *res publica* and later the republic as an instance transcendent to the agents (the king included) who are its temporary incarnations. (1994a, p. 130 [p. 58])

Thus far, Bourdieu had described the 'strategies' adopted by agents or groups within particular fields as being determined by their inherent 'interest' in conserving or accumulating the specific forms of capital on offer within the field in question. Here, he argued that the relative autonomy of a field such as the bureaucratic meant that agents could only pursue such partial 'interests' 'at the cost of a submission (if only in appearance) to the universal' (1994a, p. 131 [p. 59]). The relative autonomy of the bureaucratic field resided precisely in its ability to determine its own rule of functioning and, in this case, the field was governed by 'the interest in disinterest-edness', a principle which obliged agents to accord their particular interests with a 'universal' interest, namely the pursuit of the public good (1994a, p. 133 [p. 60]). As long as the bureaucratic field retained its relative autonomy from the fields of politics or the economy, agents could only conserve and accumulate the symbolic capital on offer within that field by respecting the public service ethos. This Bourdieu termed 'a corporatism of the universal'.

Throughout *The State Nobility* and *The Weight of the World*, Bourdieu's primary concern had, of course, been the erosion of this autonomous principle in the educational, political and bureaucratic fields as the heteronomous principles of the market gained ever greater sway. Bourdieu's understanding of the terms 'autonomy' and 'heteronomy' seemed to mirror Weber's definition in *Economy and Society*: 'An organisation may be autonomous or heteronomous. ... Autonomy means that the order governing the organisation has been established by its own members on their own authority. ... In the case of heteronomy, it has been imposed by an outside authority' (Weber 1968, pp. 49–50). Throughout his recent theoretical work and his more directly political interventions, Bourdieu has argued for the need to strengthen the autonomy of the various fields. By allowing the educational, political, intellectual and bureaucratic fields to function according to 'the order established by its own members', he maintains, those members will be forced to pursue strategies in which their particular interests accord with universal interests and hence, according to the principles of a 'corporatism of the universal', contribute to the 'advance of the universal'.

Thus, for example, in *Méditations pascaliennes*, he argued that strengthening the autonomy of the political field against the heteronomous forces of the market was one way to ensure that politicians would be forced to obey universal ethical criteria and hence work to 'advance the universal' in the political field. Writing against a background of numerous scandals involving financial corruption amongst France's ruling elite, he maintained that the autonomy of the political field rested on its distance from the market, from the realm of economic necessity, 'on the *skholè* and on the scholastic distance vis-à-vis, especially economic, necessity and urgency' (1997, p. 131). In his studies of Kabylia, as in *Distinction*, Bourdieu had identified this 'scholastic distance' on the world as not only profoundly distorting, but also the preserve of the leisured bourgeoisie. Here, however, he insisted on its inherent ambiguity as both a mark of social privilege *and* the precondition for autonomous thought and action:

> If the universal advances, it is because there exist social microcosms which, despite their intrinsic ambiguity, linked to their enclosure in privilege and the satisfied egotism of a distinctive separation, are the sites of struggles in which the universal is at stake and in which agents having ... *a particular interest in the universal*, in reason, in truth, in virtue, take up arms and fight with weapons which are nothing other than the most universal conquests of prior struggles. (1997, p. 146)

Bourdieu thus called for a reinforcement of the political field's autonomy, a '*Realpolitik* of reason' which would aim 'to install or reinforce, at the heart of the political field, mechanisms capable of imposing sanctions, as far as possible automatic sanctions, of a sort to discourage lapses in the democratic norm (such as the corruption of elected representatives) and to encourage or impose proper behaviour' (p. 150).

As Yves Sintomer points out, the problem with this conception of the political field was its unintended elitism. By Bourdieu's own admission, the 'scholastic distance' from material necessity, which he identified as the precondition for 'advancing the universal' in the political field, was the preserve of the wealthy bourgeoisie. The possibility of any genuinely participatory democracy thus seemed excluded: 'The politics in question can only be a matter for very specific individuals and does not refer to the action of ordinary citizens' (Sintomer 1996, p. 94). Through the notion of 'the corporatism of the universal', Bourdieu had clearly hoped to reconcile the potentially contradictory halves of his output, that is to say to reconcile a critique of the arbitrary and elitist nature of autonomous fields and their products with a defence of that autonomy against the incursions of the market. The unintended elitism of his account of the political field suggested that attempted reconciliation had not been completely successful. By examining Bourdieu's defence of the autonomy of the fields of artistic and intellectual production, it will be possible to see to what extent he managed to answer these criticisms.

Intellectual Autonomy in the Face of the Market

The Rules of Art represents Bourdieu's most detailed analysis to date of the way in which a particular field, in this case the field of literary and artistic production, was able historically to gain its relative autonomy from heteronomous forces of economic, political and state power. *The Rules of Art* focused on the struggles of nineteenth-century artists and writers such as Manet, Flaubert and Baudelaire to secure their creative autonomy from the courts, the Church, the *Salon* and the *Académie*. Bourdieu argued that in their struggles against *Salon* refusals and obscenity trials, Manet, Flaubert and Baudelaire not only fought for their own artistic autonomy but also contributed to the emergence of an autonomous field of artistic production. This autonomous or '*restricted* field of production' was opposed to an '*enlarged* field of production', subservient to the demands of the market. Within the restricted field of production, artists and writers produced cultural artefacts which were destined, initially at least, to be consumed not by a mass market but by an audience restricted to their peers, to other artists and a few enlightened critics. The restricted field was thus autonomous in the sense that it functioned according to rules established by its members. More specifically, the restricted field was 'an inverted economy' in which immediate 'temporal reward', whether in the form of celebrity, economic success or political favour, would be regarded with suspicion by one's fellow artists. The price of accumulating symbolic capital within the restricted field was precisely renunciation of the immediate rewards of wealth or fame on offer in the enlarged field of production. This 'dual' or 'bi-polar' structure, in which the restricted field of production was opposed to the field of enlarged production, ensured that artists and writers in the restricted field could only pursue their particular interests in that field by 'submitting to the universal', by producing works not intended primarily to secure a commercial return, but rather 'to advance the universal'.

Bourdieu's insistence in *The Rules of Art* that the products of an autonomous field of cultural production could possess a universal value represented a significant shift in his position. In an article first published in the mid-1970s, 'The Specificity of the Scientific Field', he had distinguished between an autonomous scientific field, which demanded respect for 'the universal norms of reason', and 'the religious field (or the field of literary production), in which official truth is nothing other than the legitimate imposition (i.e. arbitrary imposition, misrecognised as such) of cultural arbitrariness expressing the specific interest of the dominant' (1976a, p. 100). In *Distinction*, Bourdieu had similarly understood high culture as a purely arbitrary construct with no objective function other than the legitimation of the dominant class's social distinction. In *The State Nobility*, however, he had argued that the relative autonomy enjoyed by artists and intellectuals meant they possessed 'the potential for subversive *détournement*' of their accumulated capital to the ends of social or political critique. As the dominated fraction of the dominant class, the intelligentsia had an objective affinity with the dominated classes per se and this encouraged 'subversive alliances, capable of threatening the social order', through struggles to impose a new 'vision and division' of the social world (1989, pp. 556–7 [p. 387]).

Bourdieu's reappraisal of the 'subversive potential' of autonomous artists and their art had thus culminated, in *The Rules of Art*, in his stated desire to return to 'the heroic times' when artistic autonomy was first wrested from the power of Church, state and market. To reconstruct the emergence of the autonomous fields of literary and artistic production was also, he argued:

> to return to the 'heroic times' of the struggle for independence, when virtues of revolt and resistance had to assert themselves clearly in the face of a repression exercised in all its brutality (especially during the trials), and to rediscover the forgotten, or repudiated, principles of intellectual freedom. (1992a, p. 76 [p. 48])

Quite why Bourdieu felt the need to return to the 'heroic times' of Flaubert, Manet and Zola was made clear in a polemical postscript he appended to *The Rules of Art*, entitled simply 'For A Corporatism of the Universal'. Here he painted a gloomy picture of an intellectual and artistic field increasingly beholden to the demands of advertising, the media and commercial sponsors. Claiming that these forces conspired to promote only those artists and intellectuals who were 'media friendly', Bourdieu lambasted a whole generation of 'philosopher journalists' who might have been adept at courting the media but whose work was a caricature of genuine intellectual endeavour. Although mentioning no one intellectual by name, his references to a book and television series which had travestied the long French tradition of critical left-wing thought was clearly an allusion to Bernard-Henri Lévy's notorious *Adventures on the Freedom Road* (1991).

The figure of Bernard-Henri Lévy could also be discerned behind Bourdieu's critique in *The Weight of the World* of the contribution of certain, again unnamed intellectuals to the forging of a neo-liberal consensus and the erosion of the public service ethos:

> a demolition of the idea of public service, in which the new master thinkers have collaborated by a series of theoretical falsehoods and dubious equations ... : by making economic liberalism the necessary and sufficient condition of political liberty, state inter-ventionism is assimilated with 'totalitarianism', the struggle against inequalities, seen as inevitable, is taken as being pointless and in any case detrimental to freedom. (1993, p. 221)

Bourdieu's target here was clearly that somewhat amorphous group of intellectuals known as '*les nouveaux philosophes*', who had emerged to great media acclaim in the mid-1970s. Perhaps best personified by Lévy himself and the ex-Maoist André Glucksmann, they typically shared a past in the radical politics of the far left *groupuscules* which they had since publicly and noisily abandoned in favour of an ethical and economic liberalism now seen as an essential defence against an irredeemably totalitarian tradition of Marxist and *marxisant* thought and politics. Not only were the careers of the *nouveaux philosophes* advanced by their high media profile, they were also closely associated with Giscard's politics, with many on the Left seeing their theoretical work as part of a directly political campaign to discredit the 'Common Programme' of the 1970s, which united Communists, Socialists and Radicals around a strongly interventionist agenda.

In *Distinction*, Bourdieu had attributed the success of the *nouveaux philosophes*, of 'intellectual producers more directly subordinated to the demand of economic and political powers' to the emergence of a mass market for intellectual ideas increasingly controlled by the new media industries (1979, p. 169 [p. 152]). In *On Television and Journalism*, he extended this list of 'philosopher journalists' to include figures such Alain Finkielkraut, Alain Minc, Luc Ferry and

Alain Renaut, bemoaning the fact that they had all gained celebrity and prestige outside the traditional academic 'instances of consecration', through appearances on the French media, which now claimed to be a 'legitimate instance of legitimation' (1996, p. 28 [p. 27]). By means of the media, the heteronomous forces of the market were thus making inroads into the once autonomous field of intellectual production.

It was in *The State Nobility* that Bourdieu offered his most detailed account of the imposition of a new model of intellectual labour, subservient to the demands of political, economic, or media power. Here he argued that one of the stakes in the rise of the heteronomous ENA at the expense of the more autonomous ENS had been to redefine the intellectual as a technocratic expert and hence challenge the existence of an autonomous intellectual field structured by its own internal principles of behaviour, legitimation and reproduction. The much vaunted 'victory' or 'revenge' of the liberal Raymond Aron, 'beacon author for Sciences-po and ENA', over the Marxist existentialist Sartre was, Bourdieu argued, merely one symptom of this general phenomenon, of 'the pretension of the technocrats who, exercising a temporal power in the name of an academic guarantee, consider it increasingly within their right to exercise intellectual authority in the name of their temporal power' (1989, pp. 302–3 [pp. 212–13]).

As the symbolic prestige associated with genuinely academic, as opposed to purely technocratic labour declined alongside the relative material wealth of academics, so they began to seek compensation outside the French academic field, whether in the form of paid consultancies, visiting professorships, notably in the United States, or of success in the 'secular' fields of publishing and journalism (pp. 297–8 [pp. 209–10]). In the field of the humanities, an older culture based on its rarity value was being replaced by the more technocratic disciplines of marketing, public relations and business studies. In the field of the natural sciences, meanwhile, the increasing complexity of scientific discovery combined with concentration in the business sector to increase institutional control over intellectuals frequently working in large publicly or privately-funded research organisations. In both fields an 'intellectual artisan class' was being supplanted by 'an intellectual wage-earning class' more directly beholden to economic and political power. New forms of state patronage were emerging in the fields of artistic and scientific endeavour alike (pp. 482–6 [pp. 336–9]). Faced with this situation, Bourdieu suggested intellectuals had a clear choice: either they could accept the new definition of the intellectual as expert and technocrat or they could, 'efficiently (that is, using the weapons of science) assume the function that was for a long time fulfilled by the intellectual, that is, enter the political arena in the

name of the values and truths acquired in and through autonomy'
(p. 486 [p. 339]).

With increasing frequency, Bourdieu has cited Émile Zola as a
model for this kind of principled intellectual intervention, stating,
for example, that 'the inaugural archetype of intellectual engagement
is represented by Zola's actions at the moment of the Dreyfus
Affair' (1997a, p. 65). In *The Rules of Art*, he argued that it was
precisely in the name of the universal values guaranteed by the
autonomy of the artistic and literary fields, that Zola was able to
intervene directly in the political field of his day:

> By an apparent paradox, it is only at the end of the century, at
> a time when the literary field, the artistic field and the scientific
> field arrive at autonomy, that the most autonomous agents of
> these autonomous fields can intervene in the political fields as
> intellectuals ... , that is, with an authority founded on the
> autonomy of the field and all the values associated with it: ethical
> purity, specific expertise, etc. In a concrete fashion, intrinsically
> artistic or scientific autonomy is asserted in political acts like Zola's
> 'J'accuse' and the petitions designed to support it. (1992a,
> pp. 464–5 [p. 342])

Manet, Flaubert, Baudelaire and particularly Zola have thus
become models for Bourdieu of a tradition of principled
autonomous intellectual engagement which he considered
threatened by the developments he had traced in *The State Nobility*
and elsewhere. In both *Free Exchange* and *On Television*, he would
invoke the example of these nineteenth-century artists again,
opposing them to 'philosopher journalists' such as Lévy, Ferry,
and Finkielkraut. Zola, in particular, served Bourdieu as a model
of the autonomous intellectual:

> According to the model invented by Zola, we can and must
> intervene in the world of politics, but with our own means and
> ends. Paradoxically, it is in the name of everything that assures
> the autonomy of their universe that artists, writers or scholars
> can intervene in today's struggles. (1994, p. 38 [p. 29])

The paradox referred to here reflected Bourdieu's belief that it was
only by distancing themselves from the world that artists and
intellectuals could gain influence over the world, only by resisting
the temporal rewards of the market could they retain their critical
force. The 'grandeur of the old-style intellectual' resided, Bourdieu
argued, in their 'critical dispositions based on independence from
temporal demands and seductions' (1994, p. 59 [p. 52]). This notion
that intellectual authority derived from a renunciation of 'temporal
rewards' recalled Weber's description of the 'charisma' of religious
prophets, whose asceticism or renunciation of the world paradoxically

increased their influence over worldly matters (see Gerth and Mills, eds 1948, pp. 323–59). Bourdieu's belief that in a 'restricted field' of artistic or intellectual production and reception, the judgement of one's peers would ensure one's work met certain universal criteria, on the other hand, recalled Bachelard's notion of a 'scientific' or 'intellectual city' composed of suitably qualified scientists, whose judgement of the work of their colleagues would ensure it met communally accepted criteria of epistemological validity. Thus, in *On Television*, Bourdieu argued that 'heteronomy' had entered the intellectual field 'when someone who is not a mathematician intervenes to give an opinion about mathematics, or when someone who is not recognised as a historian (a historian who talks about history on television, for example) gives an opinion about historians – and is listened to' (1996, p. 66 [p. 57]).

As early as *The Craft of Sociology*, Bourdieu had invoked Bachelard's vision of 'a homogeneous, well-guarded scientific city' as a defence against the temptation for social scientists to be influenced by the 'worldly' concerns of intellectual fashion (1968, p. 347 [p. 233]). Citing Bachelard's analysis of the direct influence of fashion and the search for prestige within the Court over the work of eighteenth-century physicists, he had suggested that the contemporary equivalents of such physicists could be found working in 'psychoanalysis, anthropology and even sociology' (p. 103 [p. 69]). In *Homo Academicus*, Bourdieu seemed to give a clearer indication of precisely who he had in mind by associating thinkers such as Barthes, Derrida, Foucault and Deleuze with the increasing interpenetration of the fields of journalism and academia. However, as we saw in Chapter 3, Bourdieu's analysis of these thinkers' careers seemed to posit a direct causal link between the media appeal of their work and its inherent merits, as though the mere fact that it had a resonance outside academia was itself proof of its superficiality. In citing the example of Zola as a model for autonomous intellectual intervention, Bourdieu seemed now to have posited a direct causal relationship between the autonomy of the field of artistic or intellectual production, the inherent value of the works produced in such a field, *and* the political morality of their producers.

However, it was not clear whether the values advanced in the restricted field of artistic production were 'universal' in purely aesthetic terms, or in political and ethical terms also. Bourdieu seemed to suggest that participation in an autonomous field was enough in itself to ensure respect for 'norms of cognitive and ethical universality and genuinely to obtain behaviour which conforms to the logical and moral ideal' (1997, p. 146). Yet issues of cognitive or logical truth and ethical or moral probity, to say nothing of aesthetic value, can surely not be so easily conflated. Not every participant in the nineteenth-century field of restricted artistic production, for example, supported the cause of the *Dreyfusards*. Edgar Degas was a notorious anti-Semite and *anti-Dreyfusard*. Bourdieu's notion of a direct causal link between autonomy and the 'universal' not only seemed incapable of accounting for individuals like Degas or those other nineteenth-century artists who were *anti-Dreyfusards*, it also had little to say about the nature of the relationship between Degas's objectionable politics and the aesthetic value of his paintings.[9]

Furthermore, it was not immediately clear how a body of 'legitimate' works which, Bourdieu had argued in his earlier work, represented nothing more than the expression of a social elite's distanced relation to the social world and could thus only be appreciated by that elite, could now be attributed a 'universal' value. As in his description of the political field, there was an implicit and unintended elitism at work here; the 'universal' could apparently only be appreciated or advanced by those lucky enough to enjoy the benefits of a leisurely 'scholastic distance' from the realm of material necessity. Anticipating such an objection, in *On Television* Bourdieu argued that at the same time as defending the autonomy of those fields in which the 'universal' was advanced, 'we must work to generalise the conditions of access to the universal, in order that more and more people fulfil the necessary conditions for appropriating the universal for themselves' (1996, p. 77 [p. 66]). In more specific terms, he advocated a classically French republican solution to this apparent paradox, namely education:

> The founders of the French Republic in the late nineteenth century used to say something that is forgotten all too often. The goal of education is not only to learn to read, write and count in order to become a good worker, it is also to gain those tools which are indispensable if one is to be a good citizen, if one is to be in a position to understand the law, to understand and defend one's rights, to set up trade unions. ... We must work to universalise the conditions of access to the universal. (p. 77 [p. 66])

On one level, this call for a return to the founding values of French republicanism was entirely consistent with Bourdieu's championing of Zola, with his critique of the deleterious effects of neo-liberalism over the fields of education, politics, bureaucracy and culture, as well as with his typically republican, even Durkheimian faith in the liberating potential of 'universal' scientific knowledge. However, on another level, Bourdieu's evident republicanism remained highly problematic. Much of Bourdieu's earlier work had been given over to undermining the claims that French education and culture might make to the universality of their values by demonstrating the historically arbitrary, partial and class-based nature of those values. Faced with a growing neo-liberal consensus and the increasing influence of market forces over the cultural and intellectual fields, Bourdieu had turned his attention to the defence of artistic and intellectual autonomy. The notion of a 'corporatism of the universal' was clearly intended as a means to reconcile these two potentially contradictory halves of his work by grounding 'universal' artistic and intellectual values on entirely contingent, historically arbitrary bases. However, the fact that, by Bourdieu's own definition, access to the 'universal' values articulated in the restricted fields of artistic and intellectual production remained the preserve of the privileged few suggested that he had ultimately failed to square the circle and that, as Nicholas Garnham has argued, there remained an unresolved 'tension' between his theoretical work and his political practice (in Calhoun et al. 1993, p. 180).

It is arguable that some such tension has been discernible in Bourdieu's work from as early as the publication of *The Inheritors* in 1964, where it manifested itself in the apparent contradiction between a critique of the arbitrary and class-based nature of French higher education and a certain residual faith in French republican ideals and values. It might be said, then, that the political and economic conjuncture of the 1980s and 1990s had merely rendered that apparent contradiction more conspicuous. What is surely beyond doubt is the immense detail and prescience with which Bourdieu has traced the emergence of a neo-liberal consensus in France and the fact that his political interventions against that consensus have tapped into broader fears and anxieties in French society as a whole, making of Bourdieu one of the most high-profile intellectuals working in that society today.

Notes to Chapter 7

1. Much of *On Television and Journalism* was given over to transcripts of two televised lectures which Bourdieu had given on Paris Première in May 1996.
2. Bourdieu has since stated that he 'no longer feels affinities' with the International Parliament of Writers (1998, p. 8n.1 [p. viii n.2]).
3. For a useful summary of the events of Autumn 1995, as well as copies of the petition in support of the strikers drafted by Bourdieu and the rival petition which appeared in *Esprit*, see the 'dossier' in *French Politics and Society* (1996). For a critique of the CFDT's stance at the time of the strikes, see Aparicio, Pernet and Torquéo (1999).
4. The École des hautes études commerciales (HEC) was founded in 1881 by the Paris Chamber of Commerce and Industry. Since the training it offered was more vocational than academic, it was for a long time looked down upon, its students traditionally referred to as 'grocers'.
5. The École normale supérieure (ENS) was founded in 1794 to train university lecturers and researchers in sciences and the arts. Traditionally at the peak of the French academic hierarchy, it counts amongst its

alumni Henri Bergson, Léon Blum, Jean Jaurès, Jean-Paul Sartre, Maurice Merleau-Ponty, Jacques Derrida and Bourdieu himself.

6. The École polytechnique was founded in 1794 to train engineers. Run by the Ministry of Defence, its graduates have access to careers in the *grands corps* of the French civil service. The École polytechnique also has a strong reputation for scientific research.

7. 'Mines engineers' are graduates of the École des Mines, established in 1783 to train senior executives for France's burgeoning industrial sector and run by the French Ministry of Industry.

8. Although it relates to a recurrent theme in Bourdieu's work, this interview is one of those considered too specifically French to be included in the English translation. As early as *Distinction*, Bourdieu had noted the increasing prevalence of the educated middle classes amongst the personnel of all the major political parties including the *Parti socialiste*, the only notable exception being the French Communists (1979, pp. 475–6 [pp. 407–8]). In *The State Nobility*, he had noted the dominance of *énarques* over the leadership of these same parties (1989, p. 304n.43 [p. 425n.43]). Bourdieu's support for presidential candidacy in 1981 of the French comedian Coluche must also be seen in the light of his growing disillusionment with professional party politics. Although criticised as being based on populist, even Poujadist principles, Bourdieu defended his decision to support Coluche's presidential bid by describing it as 'an interested act of folly', which effectively undermined the rules of the party political game (1981, p. 7).

9. For an analysis of Degas's anti-Semitism and its relation to his aesthetic, see Nochlin (1991, pp. 141–69).

Conclusion

Arriving at a final assessment of the value of Bourdieu's immense and varied output is no easy task. By placing his work in the interrelated contexts of the intellectual field out of which it emerged and the social and cultural changes it has analysed, this study has sought to emphasise the immense perceptiveness with which Bourdieu has traced the dynamic of these changes in postwar France. Indeed, he has often shown considerable prescience, anticipating the future development of socio-cultural phenomena or crystallising a more general or vaguely perceived sense of malaise. This was particularly true of the works on French higher education he co-authored with Jean-Claude Passeron between 1964 and 1970, which both diagnosed with great acuity the travails of the French universities as they struggled to adapt to their changing role and set the agenda in the domain of the sociology of education in the years preceding and immediately following 1968.

Similarly, Bourdieu was surely one of the first commentators to grasp the contradictory effects of the postwar expansion in higher education, anticipating as early as *Distinction* the now well documented phenomenon in France whereby even the most apparently menial of jobs demands some form of post-secondary educational qualification. His analysis, in the same text, of the new forms of bourgeois and petty-bourgeois culture and manners which accompanied what he termed 'reconverted conservatism', the emergence of a dominant liberal or neo-liberal political ideology, should also be credited with great prescience. Later works such as *The State Nobility* and *The Weight of the World* traced in sharper detail the rise of this new 'dominant ideology' and the networks of educational, bureaucratic, political and economic power which sustained it.

To seek to situate Bourdieu's work firmly within its *national* context of origin, at the very time when Bourdieu himself is insisting on the need for intellectuals to form *transnational* alliances in the struggle against neo-liberal hegemony, may seem somewhat perverse. However, it might be argued that it is only by understanding the particular national and historical context out of which Bourdieu's work emerged that we can assess how useful his insights might be when applied to our own national and historical context. For instance, once it has been grasped that Bourdieu's analyses of French higher education turn upon the contradictions provoked by rapid expansion in a previously elite sector, those analyses gain an increased relevance when considering the current state of British universities. Bourdieu's descriptions of what happens when new categories of university entrant encounter a teaching body imbued with a set of cultural, intellectual and linguistic assumptions engendered under an earlier, more selective system cannot help but have a particular resonance for all those working in British universities in the wake of the rapid expansion in student numbers of the early 1990s. Similarly, when considering Bourdieu's analysis of the formation of a social, political, economic and administrative elite in *The State Nobility*, a detailed understanding of the role of the French *grandes écoles* in the formation of a 'republican elite' and of the way in which neo-liberalism has been mediated through a long French tradition of state interventionism will be essential before any attempt is made to apply Bourdieu's insights to an analysis of the equivalent role of, say, Oxford and Cambridge universities in Great Britain or Harvard, Yale and Princeton in the United States. In short, this study has attempted to suggest that understanding the context in which Bourdieu's works were written is the necessary precursor to the application of his immensely provocative insights and concepts to other contexts, other national or historical settings.

The greatest compliment one can pay to a thinker like Bourdieu is, of course, precisely to take up his ideas and concepts and attempt to apply them in new areas of enquiry. If an adequate grasp of the historical and social phenomena on which Bourdieu comments is one precondition for any such application of his thought, another is surely a critical and objective assessment of the theoretical apparatus he brings to bear on those phenomena. As this study has traced the development of Bourdieu's theoretical apparatus, a series of potential contradictions or problems inherent in some of his most central concepts have emerged.

'Doxa' and the 'Habitus'

At the heart of Bourdieu's understanding of the workings of the habitus is what, following Husserl, he terms the 'doxa' or the 'doxic relation' to the world, that pre-reflexive, pre-predicative orientation towards the future ('*l'à-venir*'), an implicit or 'practical' sense of what can and cannot be reasonably achieved, of what does or does not fall within a particular historically and culturally determined 'horizon of possibilities'. With the exception of those sporadic moments of 'crisis', during which agents' investment in the apparent self-evidence of the doxa is subjected to 'a break', or a collective '*epoche*', it is this 'doxic relation' to the social world, Bourdieu argues, that gives actions their sense of purpose and meaning, that ensures there is a time and a place for everything, that naturalises and legitimises the social roles adopted by different classes, age groups and genders.

The importance of the notion of the doxa or the doxic relation and its centrality to Bourdieu's understanding of structure and agency first became evident in the essays he wrote on the Algerian peasantry and sub-proletariat at the very beginning of his career. Its influence could also be seen behind his account of the way different classes internalised their objective chances of educational success into their different habitus. However, it was with Bourdieu's elaboration of a theory of practice in his works of Kabyle ethnology that the notion of the doxa began to take on an absolutely key role in his sociological theory. It was the doxic relation to the social world which pre-disposed not only the Kabyles to accept without question the collective rhythms of their 'pre-capitalist' universe but also French academics to acquiesce to the rigid hierarchies and long wait for promotion that characterised the 'traditional' system of higher education prior to its rapid expansion in the 1960s. Bourdieu's description of the rigidity of gender divisions in Kabylia and of their perpetuation in the West also turned on the notion that such roles were internalised at the doxic, pre-predicative level. His highly contentious assertion that the working class had no culture as such relied on the assumption that their material conditions rendered them incapable of staging that break with the realm of necessity, of doxic immediacy, which was the precondition for any aesthetic experience.

If, by this notion of the doxic relation at the heart of the habitus, Bourdieu seeks to describe the way a historical process of social, cultural, and intellectual inculcation becomes occluded so that a series of culturally arbitrary conventions and norms become naturalised and hence legitimised, then this appears to be a convincing account. The habitus, in this case, describes the process whereby a set of norms and conventions becomes sedimented into a structure of dispositions and expectations, of 'practical taxonomies', of ways of seeing and doing in the world that are neither entirely conscious nor wholly unconscious but rather 'practically' oriented towards certain implicit goals. By this definition, it is possible to see that the concepts of practice and habitus attempt, with some success, to mediate between subjectivist and objectivist modes of thought and offer a potentially powerful and persuasive tool for the analysis of social action.

However, Bourdieu also frequently implies that the internalisation of social imperatives into the habitus occurs in an immediate, mimetic way, by a process of incorporation, 'a sort of symbolic gymnastics', 'from body to body, i.e. on the hither side of words or concepts' (1977a, p. 2). Here, Bourdieu invokes the notion of doxa or the doxic relation to distinguish his approach from Marxist or *marxisant* theories of ideology, which he considers to be too rationalist, too concerned with the inculcation of *ideas* and not sensitive enough to the *incorporation of bodily dispositions*. It is because social imperatives are incorporated at this doxic level, he argues, that they cannot be effectively challenged by a mere '*prise de conscience*'; they are more profoundly rooted and hence more enduring than a conventional theory of ideology might suggest.

This emphasis on the doxic, the embodied, the immediate nature of practice and habitus risks transforming the latter from a structure that has been historically determined into one that is merely culturally arbitrary. It is here that a certain pessimism, determinism and stasis enter Bourdieu's work. The habitus may engender a range of relatively unpredictable outcomes and strategies but these will always be determined by the imperatives of social reproduction or of the conservation and accumulation of stocks of symbolic or material capital, imperatives of which agents themselves remain, by definition, unconscious. Intellectuals, such as feminists, may criticise inequalities between the sexes but these criticisms will always fall wide of the mark since they are addressed to gender as an ideological construct rather than an embodied practice. Significantly, Bourdieu never convincingly explains what kind of politics might address inequalities of class or gender at that embodied level. This study has, therefore, sought to question the adequacy, at both the empirical and the theoretical level, of the notion of the doxa or the doxic relation to the world wherever Bourdieu used it, whether to analyse the politicisation of the Algerian masses, the aesthetic practices of the French working class, or the endurance of 'male domination' in Western societies.

Of Science and Scientificity

The problematic notion of doxa has a clear impact on Bourdieu's understanding of agency, since if agents are a priori assumed to enjoy a pre-reflexive relationship to their social world then their capacity to act on that world through rational action must be considered severely limited. However, it is also intimately connected with Bourdieu's claims to the scientificity of his sociological theories. For it is precisely by dint of his ability to achieve an objective distance on social phenomena not available to agents on the ground that Bourdieu can claim the scientificity of his own work. As he argued in *The Craft of Sociology*, sociology's claim to scientificity is inseparable from what he termed 'the principle of non-consciousness', the assumption that agents on the ground remain unconscious of the objective logic of their social universe, a logic which can only be laid bare by an epistemological break which is the preserve of the social scientist: 'the principle of non-consciousness, conceived as the sine qua non for the constitution of sociological science, is nothing other than the reformulation in the logic of that science of the principle of methodological determinism which no science can reject without disowning itself as science' (1968, p. 38 [p. 16]).

Interestingly, it is precisely this founding assumption of Bourdieu's sociology that his former close collaborator Luc Boltanski has sought to challenge in his recent work. As Boltanski argues, agents are, in fact, endowed with 'critical resources and put them into effect almost all the time in the ordinary run of social life, even if their criticisms have very unequal chances of modifying the state of their surrounding world, depending on the degree of control they enjoy over their social environment' (1990, p. 54). As Boltanski's remarks make clear, to emphasise the importance of such capacities is neither to slip into a naive subjectivism nor to deny any role to the sociologist in the elucidation of the specific structural constraints on critically inspired political or social action. Rather, it is a matter of questioning the absolute distinction between the scientific knowledge of the sociologist and the doxic, pre-reflexive or purely practical knowledge which Bourdieu attributes to 'ordinary' individuals in their everyday behaviour. Moreover, as Yves Sintomer has pointed out, in its continued reliance on the opposition between science, the preserve of the detached sociological observer, and doxa, the realm of pre-reflexive immediacy inhabited by other agents, Bourdieu's sociology finds it difficult to theorise the possibility of a genuinely participatory democracy and, albeit unintentionally, implies a certain elitism (1996, p. 94).

Bourdieu's recent articles and speeches, collected in *Acts of Resistance,* suggest that in practice he has a very modest vision of the role of the critical intellectual as someone who places themselves in the service of workers, the unemployed or illegal immigrants in their struggles for justice, providing the theoretical bases for a critique of the dominant discourse. However, these directly political pronouncements seem to sit uneasily with a body of theoretical work which appears constantly to assert the inherent incapacity of dominated groups to gain any critical or even aesthetic distance on their social environment.

'Field' and 'Strategy'

Inseparable from Bourdieu's positing of 'the principle of non-consciousness' as the sine qua non of a scientific sociology is his assertion that the actions of individuals or the policies pursued by institutions can only be understood by placing them within their respective *fields*, within a structure of differential relations. As he put it in *The Craft of Sociology*:

> the principle of non-consciousness requires one to construct the system of objective relations within which individuals are located, which are expressed more adequately in the economy or morphology of groups than in the subject's opinions and declared intentions. Far from the description of individual attitudes, opinions and aspirations being able to provide the explanatory principle of the functioning of an organisation, it is an understanding of the objective logic of the organisation that leads to the principle capable of additionally explaining individual attitudes, opinions and aspirations. (1968, pp. 40–1 [p. 18])

There can be no doubting that the metaphor of the field as a magnetic force field with its poles of attraction and repulsion is a peculiarly persuasive one, particularly when applied to the artistic or intellectual domains. Indeed, Bourdieu's contention that artistic or intellectual productions can neither be attributed to the workings of an individual creative genius nor simply read as the reflection of 'external' historical or social conditions but need rather to be understood as the result, in part at least, of the mediating force of a field, with its own characteristic structure and specific historical genesis, is surely one of his most valuable insights. Similarly, his use of correspondence analysis as a means for plotting the different poles of opposition and affiliation which characterise a particular field can prove a powerful heuristic tool.

The problem arises, however, when Bourdieu seeks to claim some kind of ontological status for the fields he 'constructs' by means of correspondence analysis, arguing in *An Invitation to Reflexive Sociology*, for example, that this is an analytical tool 'whose philosophy corresponds exactly to what, in my view, the reality of the social world is' (1992, p. 72 [p. 96]). But the example of Bourdieu's construction of the 'field of power' in *The State Nobility* suggests there are good reasons for doubting whether correspondence analysis really reflects social reality quite so faithfully. For the 'field of power', as Bourdieu plots it, amounts to nothing more than a nationally selected sample of professionals and seems to overlook the complex question of the articulation of the forms of intellectual, cultural or economic power held by individuals or groups with the political power of democratic institutions, the power of the judiciary and the bureaucracy, or the increasingly important sources of supra-national power, whether in institutions such as the European Union or in multinational business corporations. In short, if correspondence analysis may offer a powerful model of reality, there is surely a danger in confusing this *model of reality* for the *reality of the model*, to use the distinction Bourdieu himself coined in his critique of Lévi-Strauss's models of kinship systems.

What Bourdieu's field theory seems to lack is any convincing account of the articulations between different national fields and sub-fields, and between those national fields and the increasingly important supranational fields, whether economic, political or cultural. Bourdieu's reluctance to specify a priori any 'transhistorical' rules governing the relations between fields is understandable. However, this has not prevented him from asserting that there is a transhistorical law governing the internal functioning of fields, namely that tendential law which demands that every participant in a field seeks, in accordance with an internalised 'interest' or 'sense of the game', to conserve or accumulate the capital on offer within that field. According to Bourdieu, it is this 'interest', this inherent tendency to accumulate capital that is the ultimate determinant of social action, of the 'strategies' adopted by agents in specific fields. If the concept of strategy implies an 'invariant' or 'transhistorical' impulse behind all human activity, it also risks reducing not only the determinants but also the significance of a particular course of action to the inherent need to accumulate or conserve a specific form of capital.

Thus, in the case of his reading of the Barthes-Picard Affair, Bourdieu could argue that the 'true principle' behind Barthes's theoretical innovations lay not in the content of his work but rather in the structural constraints of the French intellectual field at a given historical moment and in Barthes's strategic interest in accumulating the new forms of intellectual capital on offer there. The danger here was of confusing cause with effect. To argue that certain theoretical developments were destined to be well received in a particular historical conjuncture does not prove that those developments were determined by Barthes's 'interest', whether conscious or otherwise, in taking advantage of that conjuncture. Even supposing that Barthes's work had been determined by the new forms and sources of academic prestige available to him in the postwar French intellectual field, this does not necessarily say anything about the inherent strengths or weaknesses of that work.

In the case of his reading of the Barthes-Picard Affair, Bourdieu seemed to have conflated issues belonging to the realm of the sociology of knowledge, the historical conditions of possibility of the emergence of a form of thought, with judgements concerning the inherent validity or value of that form of thought. In his more recent work, with the elaboration of the notion of 'a corporatism of the universal', he seems to have conflated both these sets of questions with the supplementary question of the political morality of the thinker in question, arguing that those working in an autonomous field of intellectual or artistic production will contribute not only to the progress of 'universal' cognitive or aesthetic values, but also to the progress of 'universal' ethical values. Numerous historical examples could surely be found of artists who produced important work without working in an autonomous field of production, as defined by Bourdieu, or of artists who did work in such an autonomous field yet had particularly objectionable moral and political values.

French Republicanism and the 'Universal'

The terms in which Bourdieu has sought to defend the 'universal' value of science, art and culture in his more recent works have revealed something which critics have so far tended to ignore – his frequently ambivalent relationship to the French republican tradition. Historically, Durkheimian sociology was intimately linked to that tradition and to its high-water mark during the founding of the Third Republic, at the turn of the century. Although by no means always faithful to the letter of the Durkheimian tradition, Bourdieu's understanding of the role of a scientific sociology is certainly faithful to its spirit. His continued faith in the power of a scientific sociology to emancipate agents from the mystified vision of art and culture which keep them in their place, his qualified belief in the ability of suitably 'rationalised' forms of teaching to mitigate social inequalities, his determined defence of the universal values of scientific knowledge against postmodernist 'irrationalism' and 'nihilism' all point to the extent of Bourdieu's affinities with the Durkheimian tradition and the republican ideals and values that implies. Indeed, on one reading, the severity of his critiques of the institutions of French culture and education might be seen as a measure of his disappointment at these institutions' failure to achieve their meritocratic and democratic republican mission.

Bourdieu's relationship to French republicanism has never been straightforward. But if we are to judge by his invocations of Zola as a model for intellectual engagement, or his call – in the petition he drafted in support of the strikers in 1995 – for the 'universal achievements of the Republic' to be safeguarded, he would seem to have returned to certain typically French republican values as a touchstone in the face of 'the neo-liberal invasion'. This form of Third Republic universalism has been much criticised since its model has proved, in practice, incapable of accommodating certain particularisms of class, race, gender and sexuality. But of course, Bourdieu is well aware of this, acknowledging, in *Acts of Resistance*, the importance of challenging the 'false universalism of the West' (1998, p. 25 [p. 19]). Further, if many of Bourdieu's models for intellectual engagement seem to come from the specifically French republican tradition, his is by no means a parochial vision of intellectual activity. Much of his recent effort has been directed towards developing what he terms 'a new internationalism', setting up networks of political support and intellectual exchange on a transnational basis (1998, pp. 66–75 [pp. 60–9]). Similarly, *Acts of Resistance* shows evidence of the extent of Bourdieu's involvement in campaigns in defence of France's immigrant populations and against the spread of racism. In recent articles and in his support for the legalisation of same-sex marriage, he has shown his sensitivity to issues of sexuality (see Bourdieu 1997b). Bourdieu's interest in questions of gender inequality dates from as early as *The Inheritors* and has been reaffirmed with the publication of *La Domination masculine* in 1998.

Nonetheless, against this apparently inclusive, laudable vision of political and intellectual activity, we must place Bourdieu's insistence that only those endowed with 'scholastic distance' from the realm of material necessity can contribute to the elaboration or advance of the universal. Bourdieu's universalism, for all its admirable intentions, remains in the final instance strangely exclusive. If Bourdieu's political pronouncements have shown an exemplary sensitivity to issues of ethnicity and sexuality, the question of the influence of ethnicity and sexuality on lifestyle, social distinction and forms of domination, both symbolic and physical, remains largely untheorised in his work. The difficulties Bourdieu has encountered in reconciling his critique of the socially exclusive nature of certain autonomous forms of culture and learning with a defence of such autonomy suggest that his notion of a 'corporatism of the universal' will require further theoretical reflection if it too is not to become another example of 'the false universalism of the West'.

But to point to some of the problems raised or the questions left unanswered by Bourdieu's theoretical work is in no way to seek to minimise his intellectual achievements. As Lévi-Strauss put it in *The Raw and the Cooked*: 'The true scholar is not someone who provides the right answers, but someone who asks the right questions.' If we may at times want to question some of the answers Bourdieu has suggested to us, there can be little doubt that the questions he has asked over the course of his career and continues to ask today are the right ones.

Bibliography

Books and Articles by Pierre Bourdieu

The following includes only those books and articles consulted in the course of researching this study. For a complete bibliography of Bourdieu's work up until 1988, see Delsaut (1988). A detailed bibliography of his work up until 1994 can be found in Mörth and Fröhlich, eds (1994). In keeping with convention, Bourdieu's journal, *Actes de la recherche en sciences sociales*, will be abbreviated in the form '*ARSS*' throughout.

Bourdieu, P. (1958) *Sociologie de l'Algérie*, Paris: Presses universitaires of France, 'Que sais-je?', no. 802.

—— (1959) 'Tartuffe ou le drame de la foi et de la mauvaise foi', *Revue de la méditerranée*, vol.4–5, no. 92–3, pp. 453–8.

—— (1959a) 'La Logique interne de la civilisation algérienne traditionnelle', in *Le Sous-développement en Algérie*, Alger: Secrétariat social, pp. 40–51.

—— (1959b) 'Le Choc des civilisations', *ibid*, pp. 52–64.

—— (1960) 'Guerre et mutation sociale en Algérie', *Études méditerranéennes*, no. 7 (Spring), pp. 25–37.

—— (1961) *Sociologie de l'Algérie*, Paris: Presses Universitaires de France, 'Que sais-je?', no. 802, revised edition.

—— (1961a) 'Révolution dans la révolution', *Esprit*, no. 1 (January), pp. 27–40.

—— (1962) *The Algerians*, trans. A.C.M. Ross, Boston: Beacon Press.

—— (1962a) 'De la guerre révolutionnaire à la révolution', in *L'Algérie de demain*, ed. F. Perroux, Paris: Presses Universitaires de France, pp. 5–13.

—— (1962b) 'Les Relations entre les sexes dans la société paysanne', *Les Temps modernes*, no. 195 (August), pp. 307–31.

—— (1962c) 'Célibat et condition paysanne', *Études rurales*, no. 5–6, (April–September), pp. 32–136.

—— (1962d) 'La Hantise du chômage chez l'ouvrier algérien: prolétariat et système colonial', *Sociologie du travail*, no. 4, pp. 313–31.

—— (1962e) 'Les Sous-prolétaires algériens', *Les Temps modernes*, no. 199 (December), pp. 1030–51.

—— (1963) *Sociologie de l'Algérie*, Paris: Presses Universitaires de France, 'Que sais-je?', no. 802, third edition.

—— (1963a) *Travail et travailleurs en Algérie*, Paris, The Hague: Mouton (with A. Darbel, J. P. Rivet and C. Seibel).

—— (1963b) 'La Société traditionnelle: attitude à l'égard du temps et conduite économique', *Sociologie du travail*, no. 1, (January–March), pp. 24–44.

—— (1963c) 'Sociologues des mythologies et mythologies des sociologues', *Les Temps modernes*, no. 211 (December), pp. 998–1021 (with J. C. Passeron).

—— (1964) *Le Déracinement: la crise de l'agriculture traditionnelle en Algérie*, Paris: Éditions de Minuit (with A. Sayad).

—— (1964a) *Les Étudiants et leurs études*, Paris, The Hague: Mouton, Cahiers du centre de sociologie européenne (with J. C. Passeron).

—— (1964b) *Les Héritiers: les étudiants et la culture*, Paris: Éditions de Minuit (with J. C. Passeron).
[Translation: *The Inheritors: French Students and their Relation to Culture*, trans. R. Nice, Chicago: University of Chicago Press, 1979.]

—— (1964c) 'Paysans déracinés, bouleversements morphologiques et changements culturels en Algérie', *Études rurales*, no. 12, (January–March), pp. 56–94 (with A. Sayad).

—— (1965) *Un Art moyen: essai sur les usages sociaux de la photographie*, Paris: Éditions de Minuit, second edition (with L. Boltanski, R. Castel and J. C. Chamboredon).
[Partial translation: *Photography: a middlebrow art*, trans. S. Whiteside, Cambridge: Polity Press, 1990.]

—— (1965a) *Rapport pédagogique et communication*, Paris, The Hague: Mouton, Cahiers du centre de sociologie européenne (with J. C. Passeron and M. de Saint Martin).
[Translation: *Academic Discourse*, trans. R. Teese, Stanford: Stanford University Press, 1992]

—— (1965b) 'Le Paysan et le photographe', *Revue française de sociologie*, vol. VI, no. 2 (April–June), pp. 164–74 (with M. C. Bourdieu).

—— (1965c) 'The Sentiment of Honour in Kabyle Society', trans. P. Sherrard, in *Honour and Shame: the values of Mediterranean society*, ed. J. G. Persistany, London: Weidenfeld and Nicholson, pp. 191–241.

—— (1966) 'Différences et distinctions', in Darras, *Le Partage des bénéfices: expansion et inégalités en France*, Paris: Éditions de Minuit, pp. 117–29.

—— (1966a) 'La Fin d'un malthusianisme?', *ibid*, pp. 135–54 (with A. Darbel).

—— (1966b) 'La Transmission de l'héritage culturel', *ibid*, pp. 383–420.

—— (1966c) 'Condition de classe et position de classe', *Archives européennes de sociologie*, vol. VII, no. 2, pp. 201–23.

—— (1966d) 'L'École conservatrice, les inégalités devant l'école et devant la culture', *Revue française de sociologie*, vol. VII, no. 3 (July–September), pp. 325–47.

—— (1966e) 'Une Sociologie de l'action est-elle possible?', *Revue française de sociologie*, vol. VII, no. 4 (October–December), pp. 508–17 (with J. D. Reynaud).

—— (1966f) 'Champ intellectuel et projet créateur', *Les Temps modernes*, no. 246 (November), pp. 865–906.

—— (1966g) 'Une Étude sociologique d'actualité: les étudiants en sciences', *Revue de l'enseignement supérieur*, no. 4, pp. 199–208.

—— (1967) Postface to E. Panofsky, *Architecture gothique et pensée scholastique*, trans. P. Bourdieu, Paris: Éditions de Minuit, pp. 136–67.

—— (1967a) 'Sociology and Philosophy in France: death and resurrection of a philosophy without subject', *Social Research*, vol. XXXIV, no. 1 (Spring), pp. 162–212 (with J. C. Passeron).

—— (1968) *Le Métier de sociologue: préalables épistémologiques*, Paris: Mouton-Bordas (with J. C. Chamboredon and J. C. Passeron). [Translation: *The Craft of Sociology: epistemological preliminaries*, trans. R. Nice, New York/Berlin: Walter de Gruyter, 1991.]

—— (1968a) 'L'Examen d'une illusion', *Revue française de sociologie*, vol. IX, no. 2, pp. 227–53 (with J. C. Passeron).

—— (1968b) 'Eléments d'une théorie de la perception artistique', *Revue internationale des sciences sociales*, vol. XX, no. 4, pp. 640–64.

—— (1968c) 'Structuralism and Theory of Sociological Knowledge', trans. A. Zenotti-Karp, *Social Research*, vol. XXXV, no. 4 (Winter), pp. 681–706.

—— (1969) *L'Amour de l'art: les musées d'art européens et leur public*, second edition, revised and expanded, Paris: Éditions de Minuit (with A. Darbel and D. Schnapper). [Translation: *The Love of Art: European Art Museums and their Public*, trans. C. Beattie and N. Merriman, Cambridge: Polity Press, 1991.]

—— (1970) *La Reproduction: éléments pour une théorie du système d'enseignement*, Paris: Éditions de Minuit (with J. C. Passeron). [Translation: *Reproduction in Education, Culture and Society*, trans. R. Nice, London/Beverley Hills: Sage, 1977.]

—— (1970a) 'La Maison kabyle ou le monde renversé', in *Échanges et communications: mélanges offerts à Claude Lévi-Strauss à l'occasion de son 60e anniversaire*, Paris, The Hague: Mouton, pp. 739–58.

—— (1971) 'Champ du pouvoir, champ intellectuel et habitus de classe', *Scolies*, no. 1, pp. 7–26.

—— (1971a) 'Disposition esthétique et compétence artistique', *Les Temps modernes*, no. 295 (February), pp. 1345–78.

—— (1971b) 'Genèse et structure du champ religieux', *Revue française de sociologie*, vol. XII, no. 3, pp. 295–334.

—— (1971c) 'Le Marché des biens symboliques', *L'Année sociologique*, no. 22, pp. 49–126.

—— (1971d) 'Une Interprétation de la théorie de la religion selon Max Weber', *Archives européennes de sociologie*, vol. XII, no. 1, pp. 3–21.

—— (1972) *Esquisse d'une théorie de la pratique: précédée de trois études d'ethnologie kabyle*, Geneva: Éditions Droz.

—— (1973) 'Les Stratégies de renconversion: les classes sociales et le système d'enseignement', *Informations sur les sciences sociales*, vol. XII, no. 5 (October), pp. 61–113 (with L. Boltanski and M. de Saint Martin).

—— (1974) 'Avenir de classe et causalité du probable', *Revue française de sociologie*, vol. XV, no. 1, pp. 3–42.

—— (1975) 'Le Couturier et sa griffe: contribution à une théorie de la magie', *ARSS*, no. 1 (January), pp. 7–36 (with Y. Delsaut).

—— (1975a) 'L'Invention de la vie d'artiste', *ARSS*, no. 2 (March), pp. 65–93.

—— (1975b) 'Le Titre et le poste: rapports entre le système de production et le système de reproduction', *ARSS*, no. 2 (March), pp. 95–107 (with L. Boltanski).

—— (1976) 'La Production de l'idéologie dominante', *ARSS*, no. 2–3 (June), pp. 3–73 (with L. Boltanski).

—— (1976a) 'Le Champ scientifique', *ARSS*, no. 2–3 (June), pp. 88–104.

—— (1977) *Algérie 60: structures économiques et structures temporelles*, Paris: Éditions de Minuit.
[Modified translation: *Algeria 1960*, trans. R. Nice, Cambridge: Cambridge University Press, 1979.]

—— (1977a) *Outline of a Theory of Practice*, trans. R. Nice, Cambridge: Cambridge University Press.

—— (1978) 'Le Patronat', *ARSS*, no. 20–1 (March–April), pp. 3–82 (with M. de Saint Martin).

—— (1978a) 'Dialogue sur la poésie orale en Kabylie: un entretien avec Mouloud Mammeri', *ARSS*, no. 23 (September), pp. 51–66.

—— (1979) *La Distinction: critique sociale du jugement*, Paris: Éditions de Minuit.
[Translation: *Distinction: a social critique of the judgement of taste*, trans. R. Nice, London: Routledge & Kegan Paul, 1984.]

—— (1980) *Le Sens pratique*, Paris: Éditions de Minuit.

[Translation: *The Logic of Practice*, trans. R. Nice, Cambridge: Polity Press, 1990.]

—— (1980a) *Questions de sociologie*, Paris: Éditions de Minuit, new edition, with index, 1984.

[Translation: *Sociology in Question*, trans. R. Nice, London: Sage, 1993.]

—— (1980b) 'Le Capital social: notes provisoires', *ARSS*, no. 31 (January), pp. 2–3.

—— (1980c) 'Clou de Djeha: des contradictions linguistiques léguées par le colonisateur', *Libération*, 19–20 April, p. 13 (with D. Eribon).

—— (1980d) 'Le Mort saisit le vif: les relations entre l'histoire réifiée et l'histoire incorporée', *ARSS*, no. 32–3 (April–June), pp. 3–14.

—— (1980e) 'L'Identité et la représentation: éléments pour une réflexion critique sur l'idée de région', *ARSS*, no. 35 (November), pp. 63–72.

—— (1981) 'La Représentation politique; éléments pour une théorie du champ politique', *ARSS*, no. 36–7 (February-March), pp. 3–24.

—— (1982) *Ce que parler veur dire: l'économie des échanges linguistiques*, Paris: Fayard.

[Modified translation: *Language and Symbolic Power*, trans. G. Raymond and M. Adamson, Cambridge: Polity Press, 1991.]

—— (1982a) *Leçon sur la leçon*, Paris: Éditions de Minuit.

—— (1983) 'Les Sciences sociales et la philosophie', *ARSS*, no. 47–8 (June), pp. 45–52.

—— (1983a) 'The Philosophical Institution', in *Philosophy in France Today*, ed. A. Montefiore, Cambridge: Cambridge University Press, pp. 1–8.

—— (1983b) 'Vous avez dit "populaire"?', *ARSS*, no. 46 (March), pp. 98–105.

—— (1984) *Homo academicus*, new edition, with Afterword, Paris: Éditions de Minuit, 1988.

[Translation: *Homo Academicus*, trans. P. Collier, Cambridge: Polity Press, 1988.]

—— (1985) *Sociologie de l'Algérie*, Paris: Presses universitaires de France, 'Que sais-je?', no. 802, seventh edition.

—— (1985a) 'Du Bon Usage de l'ethnologie', *Awal, cahiers d'études berbères*, no. 1, pp. 7–29.

—— (1985b) 'Quand les Canaques prennent la parole', *ARSS*, no. 56 (March), pp. 69–83 (with A. Bensa).

—— (1986) 'La Force du droit: éléments pour une sociologie du champ juridique', *ARSS*, no. 64 (September), pp. 5–19.

—— (1987) *Choses dites*, Paris: Éditions de Minuit.

[Translation: *In Other Words*, trans. M. Adamson, Cambridge: Polity Press, 1990.]

—— (1987a) 'The Historical Genesis of a Pure Aesthetic', *Journal of Aesthetics and Art Criticism*, vol. XLVI, pp. 201–10.

—— (1987b) 'L'Institutionalisation de l'anomie', *Cahiers du Musée national d'art moderne*, no. 19–20, (June), pp. 6–19.

—— (1988) *L'Ontologie politique de Martin Heidegger*, Paris: Éditions de Minuit.

[Translation: *The Political Ontology of Martin Heidegger*, trans. P. Collier, Cambridge: Polity Press, 1991.]

—— (1988a) 'Flaubert's Point of View', translated by P. Parkhurst Ferguson, *Critical Inquiry*, vol. 14, no. 3 (Spring), pp. 539–62.

—— (1988b) 'Pour que vive une télévision publique', *Le Monde*, 19 October, p. 2.

—— (1989) *La Noblesse d'état: grandes écoles et esprit de corps*, Paris: Éditions de Minuit.

[Translation: *The State Nobility*, trans. L. C. Clough, Cambridge: Polity Press, 1996.]

—— (1989a) 'Mouloud Mammeri ou *La Colline oubliée*', *Awal*, no. 5, pp. 1–3.

—— (1989b) 'Tombeau pour une ambition', *Le Monde*, 11 May, p. 2.

—— (1990) 'La Construction du marché: le champ administratif et la production du "politique du logement"', *ARSS*, no. 81–2 (March), pp. 65–85 (with R. Christin).

—— (1990a) 'Pour une télévision publique sans publicité', *Le Monde*, 29–30 April, p. 2.

—— (1990b) 'La Domination masculine', *ARSS*, no. 84, pp. 2–31.

—— (1991) 'Un Analyseur de l'inconscient', preface to A. Sayad, *L'Immigration ou les paradoxes de l'altérité*, Brussels: Editions De Boeck-Wesmael, pp. 7–9.

—— (1992) *Réponses: pour une anthropologie réflexive*, Paris: Éditions du Seuil (with L. J. D. Wacquant).

[Translation: *An Invitation to Reflexive Sociology*, Cambridge: Polity Press, 1992.]

—— (1992a) *Les Règles de l'art: genèse et structure du champ littéraire*, Paris: Éditions du Seuil.

[Translation: *The Rules of Art: the genesis and structure of the literary field*, trans. S. Emmanuel, Cambridge: Polity Press, 1996.]

—— (1992b) 'In Conversation – Doxa and Common Life', *New Left Review*, no. 191, pp. 111–21 (with T. Eagleton).

—— (1993) *La Misère du monde*, Paris: Éditions du Seuil.

[Translation: *The Weight of the World: social suffering in contemporary society*, trans. P. P. Ferguson, Cambridge: Polity Press, 2000].

—— (1994) *Libre-échange*, Paris: Éditions du Seuil/ Les Presses du Réel (with H. Haacke).

[Translation: *Free Exchange*, Cambridge: Polity Press, 1994.]

—— (1994a) *Raisons pratiques: sur la théorie de l'action*, Paris: Éditions du Seuil.
[Translation: *Practical Reason: on the theory of action*, Cambridge: Polity Press, 1998.]
—— (1994b) 'Stratégies de reproduction et modes de domination', *ARSS*, no. 105, pp. 3–12.
—— (1995) 'La Violence symbolique', in *De l'égalité des sexes*, ed. M. de Manassein, Paris: Centre national de documentation pédagogique, pp. 83–7.
—— (1995a) 'La Cause de la science: comment l'histoire sociale des sciences sociales peut servir le progrès de ces sciences', *ARSS*, no. 106–7, pp. 3–10.
—— (1996) *Sur la télévision*, Paris: Raisons d'agir–Liber éditions.
[Translation: *On Television and Journalism*, trans. P. P. Ferguson, London: Pluto Press, 1998.]
—— (1996a) 'La Télévision peut-elle critiquer la télévision? Analyse d'un passage à l'antenne', *Le Monde diplomatique*, April, p. 20.
—— (1996b) 'Des Familles sans nom', *ARSS*, no. 113, pp. 3–5.
—— (1996c) 'Le Nouvel Opium des intellectuels: contre la "pensée Tietmeyer", un Welfare State européen', *Liber*, no. 29, p. 16.
—— (1997) *Méditations pascaliennes*, Paris: Collection Liber/Éditions du Seuil.
—— (1997a) *Les Usages sociaux de la science: pour une sociologie clinique du champ scientifique*, Paris: Institut national de la recherche agronomique.
—— (1997b) 'Le Champ économique', *ARSS*, no. 119, pp. 48–66.
—— (1997c) 'Quelques questions sur la question gay et lesbienne', *Liber*, no. 33, pp. 7–8.
—— (1998) *Contre-feux: propos pour servir à la résistance contre l'invasion néo-libérale*, Paris: Raisons d'agir–Liber éditions.
[Translation: *Acts of Resistance: against the new myths of our time*, trans. R. Nice, Cambridge: Polity Press, 1998.]
—— (1998a) *La Domination masculine*, Paris: Éditions du Seuil.
—— (1998b) 'Pour une gauche de gauche', *Le Monde*, 18 April, pp. 1 and 13.

Other Works Consulted

Abercrombie, N., Hill, S. and Turner, B. S. (1980) *The Dominant Ideology Thesis*, London: George Allen and Unwin.
Accardo, A. (1986) *Initiation à la sociologie, l'illusionnisme social: une lecture de Bourdieu*, revised edition, Bordeaux: Éditions Le Mascaret, 1991.
—— (1997) *Introduction à la sociologie critique*, Bordeaux: Éditions Le Mascaret.

Accardo, A. and Corcuff, P. (1986) *La Sociologie de Bourdieu: textes choisis et commentés*, second edition, revised and expanded, Bordeaux: Éditions Le Mascaret.

Ageron, C. R. (1968) 'Le "Mythe kabyle" et la politique kabyle (1871–1891)', in *Les Algériens musulmans et la France, (1871–1919)*, vol. one, Paris: Presses Universitaires de France, pp. 267–92.

—— (1979) *Histoire de l'Algérie contemporaine, tôme II, De l'insurrection de 1871 au déclenchement de la guerre de libération*, Paris: Presses Universitaires de France.

Alexander, J. C. (1995) 'The Reality of Reduction: The Failed Synthesis of Pierre Bourdieu', in *Fin de Siècle Social Theory: relativism, reduction, and the problem of reason*, London & New York: Verso, pp. 128–217.

Althusser, L. (1964) 'Problèmes étudiants', *La Nouvelle Critique*, no. 152, pp. 80–111.

—— (1965) *Pour Marx*, Paris: Éditions Maspero.

—— (1965a) *Lire le Capital*, Paris: Presses Universitaires de France, revised edition, 1996, (with E. Balibar, R. Establet, P. Macherey, and J. Rancière).

—— (1970) 'Idéologie et appareils idéologiques d'état', in *Positions*, Paris: Éditions Sociales, 1976, pp. 79–137.

—— (1974) *Philosophes et philosophie spontanée des savants*, Paris: Éditions Maspero.

—— (1995) *Sur la reproduction*, introduction by J. Bidet, Paris: Presses Universitaires de France/Actuel Marx Confrontation.

Amiot, M. (1972) 'À propos de *L'École capitaliste en France* de Christian Baudelot et Roger Establet', *Revue française de sociologie*, vol. 13, no. 3, pp. 413–20.

Antoine, G. and Passeron, J. C. (1966) *La Réforme de l'université*, preface by R. Aron, Paris: Calmann-Lévy.

Aparicio, J. C., Pernet, M., and Torquéo, D. (1999) *La CFDT au péril du libéral-syndicalisme*, Paris: Éditions Syllepse.

Armengaud, F., Jasser, G., and Delphy, C. (1995) 'Liberty, equality ... but most of all fraternity', *Trouble and Strife*, no. 31, pp. 43–9.

Aron, R. (1957) *La Tragédie algérienne*, Paris: Plon/Tribune Libre.

—— (1958) *L'Algérie et la république*, Paris: Plon/Tribune Libre.

—— (1968) *La Révolution introuvable*, Paris: Fayard.

Asad, T. ed. (1973) *Anthropology and the Colonial Encounter*, London: Ithaca Press.

Audétat, M. (1997) 'Quand Bourdieu allume la télévision', *L'Hebdo*, 23 January, pp. 56–8.

Austin, J. L. (1962) *How to do Things with Words*, eds J. O. Urmson and M. Sbisa, Oxford: Oxford University Press, 1971.

—— (1962a) *Sense and Sensibilia*, reconstructed from the manuscript notes by G. J. Warnock, Oxford: The Clarendon Press.

Bachelard, G. (1934) *Le Nouvel Esprit scientifique*, Paris: Quadrige/Presses Universitaires de France, seventeenth edition, 1987.

—— (1938) *La Formation de l'esprit scientifique: contribution à une psychanalyse de la connaissance*, Paris: J. Vrin, sixteenth edition, 1996.

—— (1940) *La Philosophie du non: essai d'une philosophie du nouvel esprit scientifique*, Paris: Quadrige/Presses Universitaires de France, eighth edition, 1981.

—— (1949) *Le Rationnalisme appliqué*, Paris: Quadrige/Presses Universitaires de France, sixth edition, 1986.

—— (1957) *The Poetics of Space*, trans. M. Jolas, Boston: Beacon Press, 1964.

—— (1971) *Epistémologie*, texts selected by D. Lecourt, Paris: Quadrige/Presses Universitaires de France.

Bakhtin, M. M. (1965) *Rabelais and his World*, trans. H. Iswolsky, Bloomington: Indiana University Press, 1984.

Bakhtin, M. M. /Voloshinov, V. N. (1971) *Marxism and the Philosophy of Language*, trans. L. Matejka and I. R. Titunik, Cambridge (Mass.): Harvard University Press.

Bauchet, P. (1962) *La Planification française: quinze ans d'expérience*, Paris: Éditions du Seuil.

Baudelot, C. and Establet, R. (1971) *L'École capitaliste en France*, Paris: Éditions Maspero.

Barthes, R. (1957) *Mythologies, suivi de Le Mythe aujourd'hui*, in *Oeuvres complètes, tôme I, 1942–1965*, edited and presented by Eric Marty, Paris: Éditions du Seuil, 1993, pp. 562–732.

—— (1963) *Sur Racine, ibid*, pp. 983–1106.

—— (1963a) 'Les Deux Critiques', *ibid*, pp. 1352–6.

—— (1966) *Critique et vérité*, in *Oeuvres complètes, tôme II, 1966–1973*, edited and presented by Eric Marty, Paris: Éditions du Seuil, 1995, pp. 15–51.

—— (1967) *Système de la mode, ibid*, pp. 129–401.

Baudrillard, J. (1968) *Le Système des objets*, Paris: Gallimard.

—— (1970) *La Société de consommation, ses mythes, ses structures*, preface by J. P. Mayer, Paris: Collection Folio/Idées, 1993.

—— (1972) *Pour une critique de l'économie politique du signe*, Paris: Gallimard.

—— (1973) *Le Miroir de la production, ou, l'illusion critique du matérialisme historique*, Paris: Galilée/Livre de poche, collection 'Biblio-Essais', 1994.

Bauman, Z. (1992) *Intimations of Postmodernity*, London & New York: Routledge.

Baxandall, M. (1972) *Painting and Experience in Fifteenth-Century Italy: a primer in the social history of pictorial style*, Oxford: Clarendon Press.

Becker, J. J. (1998) *Nouvelle histoire de la France contemporaine, tôme 19, Crises et alternances, 1974–1995*, with P. Ory, Paris: Éditions du Seuil.

Belleville, P. (1963) *Une nouvelle classe ouvrière*, Paris: Juillard.

Bénéton, P. and Touchard, J. (1970) 'Les Interprétations de la crise de mai–juin 1968', *Revue française de science politique*, vol. XX, no. 3, pp. 503–43.

Bennington, G. (1994) *Legislations: the politics of deconstruction*, London & New York: Verso.

Bennoune, M. (1988) *The Making of Contemporary Algeria: colonial upheavals and post-independence development*, Cambridge: Cambridge University Press.

Berger, S. (1987) 'Liberalism reborn: the new liberal synthesis in France', in *Contemporary France: a review of interdisciplinary studies*, eds J. Howorth and G. Ross, London: Frances Pinter, pp. 84–108.

Berstein, S. (1989) *La France de l'expansion, 1: la république gaullienne, 1958–1969*, Paris: Éditions du Seuil.

Berstein, S. and Rioux, J. P. (1995) *La France de l'expansion, 2: l'apogée Pompidou, 1969–1974*, Paris: Éditions du Seuil.

Bidet, J. ed. (1996) 'Autour de Pierre Bourdieu', *Actuel Marx*, no. 20.

Birnbaum, P. (1994) *Les Sommets de l'état: essai sur l'élite du pouvoir en France*, Paris: Éditions du Seuil/ Collection Points Essais.

Bisseret, N. (1968) 'La "Naissance" et le diplôme. Les processus de sélection au début des études universitaires', *Revue française de sociologie*, vol. IX, numéro spécial, 'Sociologie de l'éducation (II)', pp. 185–207.

Boltanski, L. (1982) *Les Cadres: la formation d'un groupe social*, Paris: Éditions de Minuit.

—— (1990) *L'Amour et la justice comme compétences: trois essais de sociologie de l'action*, Paris: Éditions Métailié.

Bon, F. and Schemeil, Y. (1980) 'La Rationalisation de l'inconduite: comprendre le statut du politique chez Pierre Bourdieu', *Revue française de science politique*, vol. 30, no. 6, pp. 1198–228.

Bonnewitz, P. (1997) *Premières leçons sur la sociologie de Bourdieu*, Paris: Presses Universitaires de France/Major Bac.

Boudon, R. (1969) 'La Crise universitaire française: essai de diagnostique sociologique', *Annales*, vol. 24, no. 3, pp. 738–64.

—— (1981) 'L'Intellectuel et ses publics, les singularités françaises', in *Français qui êtes-vous?*, eds. J. D. Reynaud and Y. Grafmeyer, Paris: La Documentation française, pp. 465–80.

Braudel, F. and Labrousse, E. eds (1982) *Histoire économique et sociale de la France, tôme IV: l'ère industrielle et la société d'aujourd'hui (siècle 1880–1980), troisième volume, années 1950 à nos jours*, Paris: Presses Universitaires de France.

Bürger, P. (1990) 'The Problem of Aesthetic Value', trans. S. Whiteside, in *Literary Theory Today*, eds P. Collier and H. Geyer-Ryan, Cambridge: Polity Press, pp. 23–34.

Cacérès, B. (1964) *Histoire de l'éducation populaire*, Paris: Éditions du Seuil.

Cahiers du L.A.S.A. (1988) *Lectures de Pierre Bourdieu*, 8–9, premier semestre.

Caillé, A. (1986) *Don, intérêt et désintéressement: Bourdieu, Mauss, Platon et quelques autres*, Paris: La Découverte/M.A.U.S.S.

Calhoun, C. (1996) 'A Different Poststructuralism', review of *Outline of a Theory of Practice*, *Contemporary Sociology, a Journal of Reviews*, 25, 3, pp. 302–5.

Calhoun, C., Lipuma, E. and Postone, M. eds (1993) *Bourdieu: critical perspectives*, Cambridge: Polity Press.

Capdevielle, J. and Mouriaux, R. (1988) *Mai 68: l'entre-deux de la modernité, histoire de trente ans*, Paris: Presses de la Fondation nationale des sciences politiques.

Caro, J. Y. (1980) 'La Sociologie de Pierre Bourdieu: éléments pour une théorie du champ politique', *Revue française de science politique*, vol. 30, no. 6, pp. 1171–97.

de Certeau, M. (1980) *The Practice of Everyday Life*, trans. S. Rendall, Berkeley: University of California Press, 1984.

Chaliand, G. and Minces, J. (1972) *L'Algérie indépendante: bilan d'une révolution nationale*, Paris: Éditions Maspero.

Charpentreau, J. and Kaës, R. (1962) *La Culture populaire en France*, Paris: Les Éditions ouvrières.

Chauviré, C. ed. (1995) *Critique*, numéro spécial: *Pierre Bourdieu*, nos 579–80.

Chevalier, P., Grosperrin, B. and Maillet, J. (1968) *L'Enseignement français de la Révolution à nos jours*, 2 vols, Paris and The Hague: Mouton.

Clifford, J. (1988) *The Predicament of Culture: twentieth-century ethnography, literature and art*, London: Harvard University Press.

Clifford, J. and Marcus, G. E. eds (1986) *Writing Culture: the poetics and politics of ethnography*, Berkeley: University of California Press.

Collectif 'Révoltes Logiques' (1984) *L'Empire du sociologue*, Paris: La Découverte/ Cahiers Libres no. 384.

Collier, P. (1993) '*Liber*, liberty, and literature', *French Cultural Studies*, 4, 3, pp. 291–304.

Colonna, F. ed. (1996) *Monde Arabe: Maghreb-Machrek*, 154, numéro spécial, *Algérie, la fin de l'unanisme: débats et combats des années 80 et 90*.

Compagnon, A. (1983) *La Troisième république des lettres, de Flaubert à Proust*, Paris: Éditions du Seuil.

Connell, R. W. (1983) 'The Black Box of Habit on the Wings of History: reflections on the theory of reproduction', in *Which Way is Up? Essays on Sex, Class and Culture*, London: George Allen and Unwin, pp. 140–61.

Daoud, Z. (1993) *Féminisme et politique au Maghreb: soixante ans de lutte (1930–1992)*, Paris: Maisonneuve et Larose.

Debray, R. (1978) *Modeste contribution aux discours et cérémonies du dixième anniversaire*, Paris: Éditions Maspero.

—— (1979) *Le Pouvoir intellectuel en France*, Paris: Éditions Ramsay/ Collection Folio/Essais.

—— (1997) 'Savants contre docteurs', *Le Monde*, 18 March, pp. 1 and 7.

Delsaut, Y. (1988) *Bibliographie des travaux de Pierre Bourdieu, 1958–1988*, Paris: Centre de sociologie européenne du Collège de France.

Derrida, J. (1967) *L'Écriture et la différence*, Paris: Éditions du Seuil.

—— (1967a) *De la grammatologie*, Paris: Éditions du Seuil.

—— (1975) 'Economimesis', in S. Agacinski, J. Derrida, S. Kofman, P. Lacoue-Labarthe, J. L. Nancy, B. Pautrat, *Mimesis, des articulations*, Paris: Flammarion, pp. 57–93.

—— (1978) *La Vérité en peinture*, Paris: Flammarion.

Deleuze, G. and Guattari, F. (1972) *L'Anti-oedipe: tôme I, capitalisme et schizophrénie*, Paris: Éditions de Minuit.

Descombes, V. (1979) *Le Même et l'autre: quarante-cinq ans de philosophie française*, Paris: Éditions de Minuit.

Destanne de Bernis, G. (1971) 'Les Industries industrialisantes et les options algériennes', *Revue Tiers-Monde*, vol. XII, no. 47, pp. 545–63.

Dine, P. (1994) *Images of the Algerian War: French Fiction and Film, 1954–1992*, Oxford: Oxford University Press.

Djender, M. (1992) 'La Berbérie: la Kabylie à travers l'histoire', in *Les Kabyles: éléments pour la compréhension de l'identité berbère en Algérie*, ed. Y. Tassadit, Paris: Groupement pour les droits des minorités, pp. 53–79.

Domenach, J. M. (1962) '*Les Damnés de la terre*', parts 1 and 2, *Esprit*, nos 3 and 4, pp. 454–63; 634–45.

Domenach, J. M. ed (1964) *Faire l'université: dossier pour la réforme de l'enseignement supérieur*, numéro spécial, *Esprit*, no. 328 (May–June).

Doubrovsky, S. (1967) *Pourquoi la nouvelle critique: critique et objectivité*, Paris: Mercure de France.

Douglas, M. (1981) 'Good Taste: review of Pierre Bourdieu, *Distinction*', *TLS*, 13 February, pp. 163–9, reprinted in *In the Active Voice*, London: Routledge and Kegan Paul, 1982, pp. 125–34.

Duchen, C. (1986) *Feminism in France from May '68 to Mitterrand*, London, Boston and Henley: Routledge and Kegan Paul.

—— (1994) *Women's Rights and Women's Lives in France, 1944–1968*, London and New York: Routledge.

Dufay, F. and Dufort, P. B. (1993) *Les Normaliens: de Charles Péguy à Bernard Henri-Lévy, un siècle d'histoire*, préface de R. Debray, Paris: Éditions Jean-Claude Lattès.

Dumazedier, J. (1962) *Vers une civilisation du loisir?*, Paris: Éditions du Seuil.

Durkheim, E. (1893) *The Division of Labour in Society*, trans. G. Simpson, New York: Macmillan, 1933.

—— (1894) *Les Règles de la méthode sociologique*, Paris: Flammarion, 1988.

—— (1897) *Suicide: a study in sociology*, trans. J. A. Spaulding and G. Simpson, London: Routledge and Kegan Paul, 1952.

—— (1897–98) 'Note sur la morphologie sociale', in *Journal sociologique*, introduction and commentary by J. Duvignaud, Paris: Presses Universitaires de France, 1969, pp. 181–2.

—— (1901–2) 'De quelques formes primitives de classification: contribution à l'étude des représentations collectives', *ibid*, pp. 395–461 (with M. Mauss).

—— (1903) 'Secular Morality', in *Moral Education: a study in the theory and application of the sociology of education*, trans. E. K. Wilson and H. Schnurer, New York: Free Press of Glencoe, 1961.

—— (1912) *The Elementary Forms of the Religious Life: a study in religious sociology*, trans. J. W. Swain, London: Allen and Unwin, 1915.

—— (1938) *L'Évolution pédagogique en France*, introduced by Maurice Halbwachs, second edition, Paris: Presses Universitaires de France, 1969.

—— (1956) *Education and Sociology*, New York: Free Press.

Dutheil, F (1993) 'L'Intellectuel dans la cité. Un entretien avec Pierre Bourdieu: 'Il faut restaurer la tradition de la vigilance', *Le Monde*, 5 November, p. 29.

Eagleton, T. (1991) 'From Adorno to Bourdieu', in *Ideology: an introduction*, London & New York: Verso, pp. 125–58.

Eribon, D. (1980) 'Pierre Bourdieu: la grande illusion des intellectuels', *Le Monde Dimanche*, 4 May, pp. 1 and 24.

Erickson, B. H. (1996) 'Culture, Class and Connections', *American Journal of Sociology*, vol. 102, no. 1, pp. 217–51.

Fabian, J. (1983) *Time and the Other: how anthropology makes its object*, New York: Columbia University Press.

Fanon, F. (1959) *L'An V de la révolution algérienne*, Paris: Éditions Maspero.

—— (1961) *Les Damnés de la terre*, preface by J.-P. Sartre, presentation by G. Chaliand, Paris: Gallimard, 1991.

Featherstone, M. (1991) *Consumer Culture and Postmodernism*, London: Sage.

Feraoun, M. (1962) *Journal, 1955–1962*, Paris: Éditions du Seuil.

Ferry, L. and Renaut, A. (1985) *La Pensée 68: essai sur l'antihumanisme contemporain*, Paris: Gallimard.

—— (1987) *68–86, Itinéraires de l'individu*, Paris: Gallimard.

Fields, A. B. (1970) *Student Politics in France: a study of the Union Nationale des Étudiants de France*, New York and London: Basic Books.

Flaubert, G. (1869) *L'Éducation sentimentale*, Paris: Gallimard/ Collection Folio, 1965.

Flynn, G. ed. (1995) *Remaking the Hexagon: the new France in the new Europe*, Boulder, San Francisco & London: Westview Press.

Forbes, J. and Kelly, M. eds (1993) *French Cultural Studies*, special number, *Pierre Bourdieu*, 4, 3.

Foucault, M. (1975) *Surveiller et punir: naissance de la prison*, Paris: Gallimard.

Fourastié, J. (1963) *Le Grand Espoir du XXe siècle*, Paris: Gallimard.

—— (1979) *Les Trente glorieuses ou la révolution invisible de 1946 à 1975*, Paris: Fayard.

Fourastié, J. and Courthéoux, J. P. (1963) *La Planification économique en France*, Paris: Presses Universitaires de France.

Fowler, B. (1997) *Pierre Bourdieu and Cultural Theory: critical investigations*, London, Thousand Oaks, New Delhi: Sage.

Frears, J. R. (1981) *France in the Giscard Presidency*, London: George Allen and Unwin.

Free, A. (1996) 'The Anthropology of Pierre Bourdieu: a reconsideration', *Critique of Anthropology*, vol. 16, no. 4, pp. 395–416.

French Politics and Society (1996) 'Debate: the movements of autumn – something new or déjà vu?', vol. 14, no. 1, pp. 1–27.

Frow, J. (1987) 'Accounting for Tastes: some problems in Bourdieu's sociology of culture', *Cultural Studies*, vol. 1, no. 1, pp. 59–73.

Garnham, N. (1986) 'Extended Review: Bourdieu's *Distinction*', *The Sociological Review*, no. 2, pp. 423–33.

Garnham, N. ed. (1980) *Media, Culture and Society*, special number on Pierre Bourdieu, vol. 2, no. 3.

Gartman, D. (1991) 'Culture as Class Symbolisation or Mass Reification? A Critique of Bourdieu's *Distinction*', *American Journal of Sociology*, vol. 97, no. 2, pp. 421–47.

Geertz, C. (1995) *After the Fact: two countries, four decades, one anthropologist*, Cambridge (Mass.): Harvard University Press.

Géhin, E. and Herpin, N. (1980) 'Deux comptes rendus de P. Bourdieu, *La Distinction: critique sociale du jugement*', *Revue française de sociologie*, vol. XXI, no. 3, pp. 439–48.

Geldof, K. (1997) 'Authority, Reading, Reflexivity: Pierre Bourdieu and the aesthetic judgement of Kant', *Diacritics*, vol. 27, no. 1, pp. 20–43.

Gerth, L. and Mills, C.W. eds (1948) *From Max Weber: essays in sociology*, Oxford: Oxford University Press.

Giddens, A. (1986) 'The Politics of Taste', review of *Distinction*, *The Partisan Review*, no. 53, pp. 300.

Giroux, H. (1982) 'Power and Resistance in the New Sociology of Education: beyond theories of social and cultural reproduction', *Curriculum Perspectives*, vol. 2, no. 3, pp. 1–13.

Goblot, E. (1925) *La Barrière et le niveau: étude sociologique sur la bourgeoisie française*, preface by G. Balandier, Paris: Presses universitaires de France, new edition, 1967.

Godin, E. (1996) 'Le Néo-libéralisme à la française: une exception?', *Modern and Contemporary France*, NS. 4, 1, pp. 61–70.

Gordon, D.C. (1968) *Women of Algeria: an essay on change*, Cambridge (Mass.): Harvard University Press.

Gorz, A. (1964) *Stratégie ouvrière et néocapitalisme*, Paris: Éditions du Seuil.

—— (1967) 'Étudiants et ouvriers', in *Le Socialisme difficile*, Paris: Éditions du Seuil, pp. 47–67.

—— (1970) 'Détruire l'université', *Les Temps modernes*, no. 285, pp. 1553–8.

—— (1980) *Adieu au prolétariat*, Paris: Éditions du Seuil.

Gramsci, A. (1966) *Il Materialismo storico e la filosofia di Benedetto Croce, Quaderni del carcere, 1*, Turin: Einaudi.

—— (1966a) *Gli Intellettuali e l'organizzazione della cultura, Quaderni del carcere, 2*, Turin: Einaudi.

—— (1966b) *Note sul Machiavelli, sulla politica e sullo stato moderno, Quaderni del carcere, 4*, Turin: Einaudi.

Grignon, C. and Passeron, J.C. (1989) *Le Savant et le populaire: misérabilisme et populisme en sociologie et en littérature*, Paris: Hautes études/Gallimard/Éditions du Seuil.

Griset, A. and Kravetz, M. (1965) 'De l'Algérie à la réforme Fouchet: critique du syndicalisme étudiant', (I) & (II), *Les Temps modernes*, no. 227, pp. 1880–902, & no. 228, pp. 2066–90.

Guillory, J. (1993) *Cultural Capital: the problem of literary canon formation*, Chicago: University of Chicago Press.

Halbwachs, M. (1912) *La Classe ouvrière et les niveaux de vie: recherches sur la hiérarchie des besoins dans les sociétés industrielles*, Paris, London, New York: Réimpressions Gordon and Breach, 1970.

—— (1955) *The Psychology of Social Class*, trans. C. Delaveney, London: Heinemann, 1958.

—— (1972) *Classes sociales et morphologie*, presented by V. Karady, Paris: Éditions de Minuit.

Halimi, S. (1997) *Les Nouveaux Chiens de garde*, Paris: Raisons d'agir–Liber éditions.

Hall, P. A. (1994) 'The State and the Market', in *Developments in French Politics*, eds P. A. Hall, J. Hayward, and H. Machin, revised edition, London: Macmillan, pp. 171–87.

Hall, S. (1977) 'The Hinterland of Science: ideology and the "sociology of knowledge"', in *On Ideology*, Centre for Contemporary Cultural Studies, London: Hutchinson, pp. 9–32.

Halsey, A. H., Heath, A. F., and Ridge, J. M. (1980) *Origins and Destinations: Family, Class and Education in Modern Britain*, Oxford: Clarendon Press.

Hamon, H. and Rotman, P. (1984) *La Deuxième Gauche: histoire intellectuelle et politique de la CFDT*, Paris: Éditions du Seuil/Points.

Hargreaves, A. G. and Heffernan, M.J. eds (1993) *French and Algerian Identities from Colonial Times to the Present*, Lewiston, Queenston, Lampeter: The Edwin Mellen Press.

Harker, R., Mahar, C. and Wilkes, C. eds (1990) *An Introduction to the Work of Pierre Bourdieu: the practice of theory*, London: Macmillan.

Harvey, D. (1989) *The Condition of Postmodernity: an enquiry into the origins of cultural change*, Oxford: Blackwell.

Hayward, J. (1986) *The State and the Market: industrial patriotism and economic intervention in France*, Brighton: Wheatsheaf Books.

Heidegger, M. (1949) *Chemins qui ne mènent nulle part*, trans. W. Brokmeier, new edition, Paris: Gallimard, 1962.

—— (1959) *An Introduction to Metaphysics*, trans. R. Manheim, New Haven and London: Yale University Press.

—— (1984) *Early Greek Thinking*, trans. D. Farrell Krell and F. A. Capuzzi, San Francisco, Harper and Row.

Hellman, J. (1981) *Emmanuel Mounier and the New Catholic Left, 1930–1950*, Toronto: Toronto University Press.

Héran, F. (1987) 'La Seconde nature et l'habitus: tradition philosophique et sens commun dans le langage sociologique', *Revue française de sociologie*, XXVIII, pp. 385–416.

Herskovits, M. J. (1938) *Acculturation: the study of culture contact*, New York: Augustin.

Herzfeld, M. (1987) *Anthropology Through the Looking-Glass: critical ethnography in the margins of Europe*, Cambridge: Cambridge University Press.

Hindess, B. and Hirst, P. Q. (1975) *Pre-capitalist Modes of Production*, London and Boston: Routledge and Kegan Paul.

Honneth, A. (1986) 'The Fragmented World of Symbolic Forms: reflections on Pierre Bourdieu's sociology of culture', trans. T. Talbot, *Theory, Culture and Society*, vol. 3, no. 3, pp. 55–67.

Honneth, A., Kocyba, H. and Schwibs, B. (1986) 'The Struggle for Symbolic Order: an interview with Pierre Bourdieu', *Theory, Culture and Society*, vol. 3, no. 3, pp. 35–51.

Horne, A. (1977) *A Savage War of Peace: Algeria 1954–1962*, London: Macmillan.

Howorth, J. and Cerny, P. G. eds (1981) *Elites in France: origins, reproduction and power*, London: Frances Pinter.

Husserl, E. (1913) *Idées directrices pour une phénoménologie et une philosophie phénoménologique pure, tôme premier, introduction générale à la phénoménologie pure*, trans. P. Ricoeur, Paris: Gallimard, 1950.

—— (1948) *Experience and Judgement: investigations in a genealogy of logic*, revised and edited by L. Landgrebe, trans. J. S. Churchill and K. Ameriks, London: Routledge and Kegan Paul, 1973.

—— (1952) *Ideas Pertaining to a Pure Phenomenology and to a Phenomenological Philosophy, second book, studies in the phenomenology of constitution*, trans. R. Rojecwicz and A. Schuwer, London: Kluwer Academic Publishers, 1989.

—— (1954) *The Crisis of European Sciences and Transcendental Phenomenology*, trans. D. Carr, Evanston: Northwestern University Press.

Jacoby, R. (1977) *Social Amnesia: a critique of conformist psychology from Adler to Laing*, Brighton: Harvester Press.

Jameson, F. (1991) *Postmodernism, or, the Cultural Logic of Late Capitalism*, London & New York: Verso.

Jenkins, R (1982) 'Pierre Bourdieu and the Reproduction of Determinism', *Sociology*, vol. 16, no. 2, pp. 270–81.

—— (1986) Review of *Distinction*, *Sociology*, no. 20, pp. 103–5.

—— (1992) *Pierre Bourdieu*, London: Routledge.

Julien, C. A. (1964) *Histoire de l'Algérie contemporaine, tôme premier, la conquête et les débuts de la colonisation (1827–1871)*, Paris: Presses Universitaires de France.

Kant, I. (1790) *Critique of Judgement*, translated with analytical indexes by J. C. Meredith, Oxford: Oxford University Press, thirteenth impression, 1991.

Khatibi, A. (1983) *Maghreb pluriel*, Paris: Denoël.

Kravetz, M. (1964) 'Naissance d'un syndicalisme étudiant', *Les Temps modernes*, 213, pp. 1447–75.

Kuisel, R. F. (1981) *Capitalism and the State in Modern France*, Cambridge: Cambridge University Press.

Lacoste-Dujardin, C. (1976) *Un Village algérien: structures et évolution récente*, Alger: SNED.

—— (1976a) 'A Propos de Pierre Bourdieu et de *l'Esquisse d'une théorie de la pratique*', *Hérodote*, 2, pp. 103–16.

—— (1997) *Opération 'Oiseau bleu': des Kabyles, des ethnologues et la guerre en Algérie*, Paris: Éditions La Découverte.

Lamont, M. (1992) *Money, Morals and Manners: the culture of the French and American upper middle class*, Chicago: Chicago University Press.

Laroui, A. (1970) *L'Histoire du Maghreb: un essai de synthèse*, Paris: Éditions Maspero.

Lash, S. (1990) *Sociology of Postmodernism*, London: Routledge.

Lecourt, D. (1974) *Bachelard ou le jour et la nuit: un essai du matérialisme dialectique*, Paris: Grasset.

Le Doeuff, M. (1987) 'Ants and Women, or, Philosophy without Borders', in *Contemporary French Philosophy*, ed. A. Phillips Griffiths, Cambridge: Cambridge University Press, pp. 41–54.

Lefebvre, H. (1962) *Introduction à la modernité, préludes*, Paris: Éditions de Minuit.

—— (1967) *Position: contre les technocrates*, Paris: Éditions Gonthier.

—— (1968) *Du droit à la ville*, Paris: Éditions Anthropos.

—— (1968a) *L'Irruption: de Nanterre au sommet*, Paris: Éditions Anthropos.

—— (1970) *Du rural à l'urbain*, texts collected by M. Gaviria, Paris: Éditions Anthropos.

Leiris, M. (1950) 'L'Ethnographe devant le colonialisme', in *Brisées*, Paris: Mercure de France, 1966, pp. 125–45.

Lenin, V. I. (1899) *The Development of Capitalism in Russia: the process of the formation of a market for large-scale industry*, in *Collected Works*, vol. 3, London: Lawrence and Wishart, 1960.

—— (1908) *The Agrarian Programme of Social Democracy in the First Russian Revolution, 1905–1907*, in *Collected Works*, vol. 13, London: Lawrence and Wishart, 1962, pp. 217–431.

—— (1918) *The State and Revolution: the Marxist theory of the state and the tasks of the proletariat in the revolution*, in *Collected Works*, vol. 25, London: Lawrence and Wishart, 1964, pp. 383–492.

Lévi-Strauss, C. (1955) *Tristes tropiques*, Paris: Plon.

—— (1958) *Anthropologie structurale*, Paris: Plon.

—— (1962) *La Pensée sauvage*, Paris: Plon.

—— (1964) *The Raw and the Cooked*, trans. J. and D. Weightman, London: Cape, 1970.

—— (1973) *Anthropologie structurale deux*, Paris: Plon.

Loesberg, J. (1993) 'Bourdieu and the Sociology of Aesthetics', *English Literary History*, vol. 60, no. 4, pp. 1033–56.

Looseley, D. L. (1995) *The Politics of Fun: cultural policy and debate in contemporary France*, Oxford and Washington D.C.: Berg.

Lorcin, P. (1995) *Imperial Identities: stereotyping, prejudice and race in colonial Algeria*, London and New York: I.B.Tauris.

Lyotard, J. F. (1974) *Économie libidinale*, Paris: Éditions de Minuit.

—— (1979) *La Condition postmoderne: rapport sur le savoir*, Paris: Éditions de Minuit.

—— (1988) *Le Postmoderne expliqué aux enfants: correspondance 1982–1985*, Paris: Éditions Galilée.

—— (1989) *La Guerre des Algériens, écrits 1956–1963*, selected and presented by M. Ramdani, Paris: Éditions Galilée

Lukes, S. (1973) *Emile Durkheim, his Life and Work: a historical and critical study*, Harmondsworth: Penguin.

McAllester-Jones, M. (1991) *Gaston Bachelard, Subversive Humanist, texts and readings*, Wisconsin: University of Wisconsin Press.

Mallet, S. (1962) *Les Paysans contre le passé*, Paris: Éditions du Seuil.

—— (1963) *La Nouvelle Classe ouvrière*, Paris: Éditions du Seuil.

de Man, P. (1971) 'Criticism and Crisis', in *Blindness and Insight: essays in the rhetoric of contemporary criticism*, New York: Oxford University Press, pp. 3–19.

Mandel, E. (1975) *Late Capitalism*, trans. J. De Bres, London: New Left Books.

Marcus, G. E. (1998) *Ethnography through Thick and Thin*, Princeton: Princeton University Press.

Marx, K. (1852) *The Eighteenth Brumaire of Louis Bonaparte*, in *Karl Marx and Frederic Engels, Collected Works*, vol. 11, London: Lawrence and Wishart, 1979, pp. 99–197.

—— (1857) *A Contribution to the Critique of Political Economy*, trans. S. W. Ryazanskaya, Moscow: Progress Publishers, 1970.

—— (1867) *Capital: a critique of political economy, vol. 1, book 1, the process of production of capital*, in *Karl Marx and Frederic Engels, Collected Works*, vol. 35, London: Lawrence and Wishart, 1996.

Marx, K. and Engels, F. (1848) *The Communist Manifesto*, with an introduction by A. J. P. Taylor, Harmondsworth: Penguin, 1967.

Mauss, M. (1950) *Sociologie et anthropologie, précédé d'une Introduction à l'oeuvre de Marcel Mauss par Claude Lévi-Strauss*, sixth edition, Paris: Presses Universitaires de France, 1978.

Mendras, H. (1967) *La Fin des paysans: innovation et changement dans l'agriculture française*, Paris: S.E.D.E.I.S.

Merleau-Ponty, M. (1942) *The Structure of Behaviour*, trans. A. L. Fisher, London: Methuen, 1965.

—— (1945) *Phenomenology of Perception*, trans. C. Smith, London: Routledge and Kegan Paul, 1962.

—— (1955) *Les Aventures de la dialectique*, Paris: Gallimard.

—— (1960) *Signs*, trans. R. C. McCleary, Northwestern University Press, 1964.

Michard-Marchal, C. and Ribery, C. (1982) 'Pierre Bourdieu: "Questions de politique"', in *Sexisme et sciences humaines: pratique linguistique du rapport de sexage*, Lille: Presses Universitaires de Lille, pp. 161–72.

Moi, T. (1991) 'Appropriating Bourdieu: Feminist Theory and Pierre Bourdieu's Sociology of Culture', *New Literary History*, vol. 22, no. 4, pp. 1017–49.

—— (1994) *Simone de Beauvoir: the making of an intellectual woman*, Oxford and Cambridge (Mass.): Blackwell.

Moi, T. ed. (1997) *Modern Language Quarterly*, special issue, *Pierre Bourdieu and Literary History*, vol. 58, no. 4.

Morin, E. (1959) *Autocritique*, Paris: Éditions du Seuil.

—— (1962) *L'Esprit du temps*, third edition, Paris: Grasset/ Livre de poche 'Biblio Essais', 1991.

—— (1967) *Commune en France: la métamorphose de Plodémet*, Paris: Fayard.

—— (1969) 'Culture adolescente et révolte étudiante', *Annales*, vol. 24, no. 3 (May–June), pp. 765–76.

Morin, E., Lefort, C. and Coudray, J. C. (1968) *Mai 68: la brèche, premières réflexions sur les événements*, Paris: Fayard.

Mörth, I. and Fröhlich, G. eds (1994) *Das symbolische Kapital der Lebensstile: zur Kultursoziologie der Moderne nach Pierre Bourdieu*, Frankfurt am Main, New York: Campus Verlag.

Mouriaux, R. and Subileau, F. (1996) 'Les Grèves françaises de l'automne 1995: défense des acquis ou mouvement social?', *Modern and Contemporary France*, N.S. 4, 3, pp. 299–306.

Nania, G. (1966) *Un Parti de la gauche: le PSU*, Paris: Gedalge.

Nizan, P. (1960) *Aden Arabie*, foreword by J.-P. Sartre, Paris: Éditions Maspero.

Nghe, N. (1963) 'Frantz Fanon et les problèmes de l'indépendance', *La Pensée*, 107, pp. 23–36.

Nochlin, L. (1991) *The Politics of Vision: essays on nineteenth-century art and society*, London: Thames and Hudson.

Ortiz, F. (1947) *Cuban Counterpoint: tobacco and sugar*, trans. H. de Onis, introduction B. Malinowski, new introduction F. Coronil, Durham (NC) & London: Duke University Press, 1995.

Ouerdane, A. (1990) *La Question berbère dans le mouvement algérien, 1926–1980*, preface by K. Yacine, Quebec: Éditions du Septentrion.

Peyre, H. ed. (1965) *Gustave Lanson: essais de méthode, de critique et d'histoire littéraire*, Paris: Hachette.

Picard, R. (1965) *Nouvelle critique ou nouvelle imposture*, Paris: Pauvert.

Plato (1930) *The Republic*, 2 vols, with an English translation by P. Shorey, London: Heinemann/Loeb Classical Library.

Poster, M. (1975) *Existential Marxism in Postwar France: from Sartre to Althusser*, Princeton: Princeton University Press.

Poulantzas, N. (1968) *Political Power and Social Classes*, trans. T. O'Hagan, London: Verso, 1978.

—— (1974) *Les Classes sociales dans le capitalisme aujourd'hui*, Paris: Éditions du Seuil.

Rancière, J. (1974) *La Leçon d'Althusser*, Paris: Gallimard/Collection Idées.

Reed-Danahay, D. (1995) 'The Kabyle and the French: occidentalism in Bourdieu's theory of practice', in *Occidentalism: images of the West*, ed. J. G. Carrier, Oxford: Clarendon Press, pp. 61–84.

Rémond, A. (1997) 'Le Petit Livre qui fâche tout rouge', *Télérama*, 5 February, pp. 52–4.

Rémond, R. (1982) *Les Droites en France*, Paris: Éditions Aubier Montaigne.

Ricoeur, P. (1969) 'Structure et herméneutique', in *Le Conflit des interprétations*, Paris: Éditions du Seuil, pp. 31–63.

Rigby, B. (1989) 'The Reconstruction of Culture: *Peuple et Culture* and the Popular Education Movement', in *The Culture of Reconstruction: European literature, thought and film*, ed. N. Hewitt, London: MacMillan, pp. 140–52.

—— (1991) *Popular Culture in Modern France: a study of cultural discourse*, London and New York: Routledge.

Robbins, D. (1991) *The Work of Pierre Bourdieu: Recognising Society*, Buckingham: Open University Press.

Roberts, H. (1981) *Algerian Socialism and the Kabyle Question*, Norwich: Monographs in Development Studies no. 8, School of Development Studies, University of East Anglia.

Roos, J. M. (1996) *Early Impressionism and the French State (1866–1874)*, Cambridge: Cambridge University Press.

Ross, K. (1996) *Fast Cars, Clean Bodies: decolonisation and the reordering of French culture*, London and Cambridge (Mass.): The MIT Press.

Said, E. W. (1978) *Orientalism*, second edition with new afterword, London: Penguin Books, 1995.

de Saint Martin, M. (1968) 'Les Facteurs de l'élimination et de la sélection différentielles dans les études de sciences', *Revue française de sociologie*, vol. IX, special issue, 'Sociologie de l'éducation (II)', pp. 167–84.

Sartre, J.-P. (1960) *The Problem of Method*, trans. H. E. Barnes, London: Methuen, 1963.

—— (1964) *Situations V: colonialisme et néo-colonialisme*, Paris: Gallimard.

Schmidt, V. A. (1996) 'Business, the State, and the End of Dirigisme', in J. T. S. Keeler and M. A. Schain, eds, *Chirac's Challenge: liberalization, europeanization, and malaise in France*, London: Macmillan, pp. 105–42.

Schnapp, A. and Vidal-Naquet, P. (1969) *Journal de la commune étudiante: textes et documents, novembre 1967–juin 1968*, Paris: Éditions du Seuil.

Schneidermann, D. (1996) 'Réponse à Pierre Bourdieu: La Télévision peut-elle critiquer la télévision?', *Le Monde diplomatique*, May, p. 25.

—— (1999) *Du journalisme après Bourdieu*, Paris: Fayard.

Shusterman, R. (1992) *Pragmatist Aesthetics: living beauty, rethinking art*, Oxford and Cambridge (Mass.): Blackwell.

Shusterman, R. ed. (1999) *Bourdieu: a critical reader*, Oxford: Blackwell.

Sintomer, Y. (1996) 'Le Corporatisme de l'universel et la cité', *Actuel Marx*, no. 20, *Autour de Pierre Bourdieu*, pp. 91–104.

Smith, R. C. (1982) *Gaston Bachelard*, Boston: Twayne Publishers.

Snyders, G. (1976) *École, classe et lutte des classes: une relecture critique de Baudelot-Establet, Bourdieu-Passeron et Illich*, Paris: Presses Universitaires de France.

Sorum, P. C. (1977) *Intellectuals and Decolonisation in France*, Chapel Hill: University of North Carolina Press.

Spivak, G. (1988) *In Other Worlds: essays in cultural politics*, New York & London: Routledge.

Stafford, A. (1997) '"Dégel", Hegel and the launch of *Arguments*', *French Cultural Studies*, 3, 24, pp. 283–94.

Stora, B. (1992) *La Gangrène et l'oubli: la mémoire de la guerre d'Algérie*, Paris: La Découverte.

Stevens, A (1981) 'The Contribution of the École Normale d'administration to French Political Life', in *Elites in France: origins, reproduction and power*, eds J. Howorth and G. Cerny, London: Frances Pinter, pp. 134–50.

Suleiman, E. (1974) *Politics, Power, and Bureaucracy in France: the administrative elite*, Princeton: Princeton University Press.

—— (1978) *Elites in French Society: the politics of survival*, Princeton: Princeton University Press.

Swartz, D. (1997) *Culture and Power: the sociology of Pierre Bourdieu*, Chicago & London: University of Chicago Press.

Talbott, J. (1981) *The War Without a Name: France in Algeria, 1954–1962*, London & Boston: Faber and Faber.

Thomas, N. (1994) *Colonialism's Culture: anthropology, travel, government*, Cambridge: Polity Press.

Thompson, J. B. (1984) *Studies in the Theory of Ideology*, Cambridge: Polity Press.

Tillion, G. (1957) *L'Algérie en 1957*, Paris: Éditions de Minuit.

Tlemcani, R. (1986) *State and Revolution in Algeria*, Boulder: Westview Press.

Touraine, A. (1965) *Sociologie de l'action*, Paris: Éditions du Seuil.

—— (1968) *Le Mouvement de mai ou le communisme utopique*, Paris: Éditions du Seuil.

—— (1969) *La Société post-industrielle*, Paris: Denoël.

Touraine, A., Dubet, F., Lapeyronnie, D., Khosrokhavar, F. and Wieviorka, M. (1996) *Le Grand Refus: réflexions sur la grève de décembre 1995*, Paris: Fayard.

Turkle, S. (1978) *Psychoanalytic Politics: Freud's French Revolution*, London: Basic Books.

Veblen, T. (1912) *The Theory of the Leisure Class: an economic study of institutions*, new edition, London: George Allen and Unwin, 1924.

Vermot-Gauchy, M. (1965) *L'Education nationale dans la France de 1975*, Monaco: Éditions du rocher/Futuribles.

Waldenfels, B. (1983) 'The Despised Doxa: Husserl and the continuing crisis of western reason', in *Husserl and Contemporary Thought*, ed. J. Sallis, London: Humanities Press, pp. 21–38.

Weber, M. (1904–5) *The Protestant Ethic and the Spirit of Capitalism*, trans. T. Parsons, London & New York: Routledge, 1992.

—— (1949) *The Methodology of the Social Sciences*, trans. and eds by E. A. Shils and H. A. Finch, Glencoe: The Free Press.

—— (1968) *Economy and Society: an outline of interpretive sociology*, 3 vols, New York: Bedminster Press, 1968.

Willis, P. (1983) 'Cultural Production and Theories of Reproduction', in *Race, Class and Education*, eds L. Barton and S. Walker, London: Croom Helm.

Wilson, E. (1988) 'Picasso and *pâté de foie gras*: Pierre Bourdieu's sociology of culture', *Diacritics*, vol. 18, no. 2, pp. 47–60.

Winock, M. (1975) *"Esprit": Des intellectuels dans la cité, 1930–1950*, Paris: Éditions du Seuil.

Wolff, J. (1983) *Aesthetics and the Sociology of Art*, London: George Allen & Unwin, second edition 1993.

Yacine, T. ed. (1992) *Les Kabyles: éléments pour la compréhension de l'identité berbère en Algérie*, Paris: Groupement pour les droits des minorités.

Young, R. (1990) *White Mythologies: writing, history and the West*, London: Routledge.

Zaoui, S. (1997) 'The Terror that Stalks the Mountains', *Index on Censorship*, 3, pp. 40–52.

Zola, E. (1880) *Le Roman expérimental*, Paris: Garnier-Flammarion, 1971.

Index